NOTICE.

Engineers and Superintendents of Gas Works will confer a favor on HARRIS & BRO., by acknowledging the receipt of the "Pocket Companion."

GAS SUPERINTENDENT'S

POCKET COMPANION

FOR THE YEAR 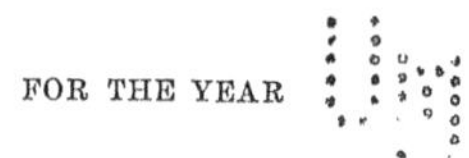

1866.

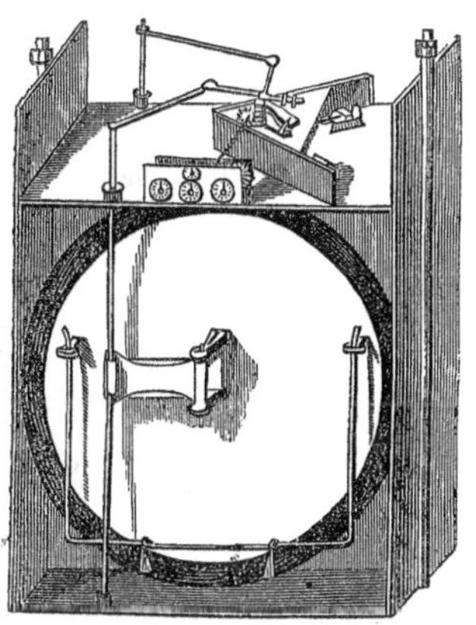

COMPILED BY DR. WM. H. M'FADDEN,

OF

HARRIS & BROTHER,

GAS METER MANUFACTURERS,

No. 1117 CHERRY STREET,

PHILADELPHIA.

ALMANAC OF 1866.

1866.	SUNDAY.	MONDAY.	TUESDAY.	WEDNESDAY.	THURSDAY.	FRIDAY.	SATURDAY.	1866.	SUNDAY.	MONDAY.	TUESDAY.	WEDNESDAY.	THURSDAY.	FRIDAY.	SATURDAY.
JAN.		1	2	3	4	5	6	**JULY.**	1	2	3	4	5	6	7
	7	8	9	10	11	12	13		8	9	10	11	12	13	14
	14	15	16	17	18	19	20		15	16	17	18	19	20	21
	21	22	23	24	25	26	27		22	23	24	25	26	27	28
	28	29	30	31					29	30	31				
FEB.					1	2	3	**AUG.**				1	2	3	4
	4	5	6	7	8	9	10		5	6	7	8	9	10	11
	11	12	13	14	15	16	17		12	13	14	15	16	17	18
	18	19	20	21	22	23	24		19	20	21	22	23	24	25
	25	26	27	28					26	27	28	29	30	31	
MAR.					1	2	3	**SEPT.**							1
	4	5	6	7	8	9	10		2	3	4	5	6	7	8
	11	12	13	14	15	16	17		9	10	11	12	13	14	15
	18	19	20	21	22	23	24		16	17	18	19	20	21	22
	25	26	27	28	29	30	31		23	24	25	26	27	28	29
									30						
APRIL.	1	2	3	4	5	6	7	**OCT.**		1	2	3	4	5	6
	8	9	10	11	12	13	14		7	8	9	10	11	12	13
	15	16	17	18	19	20	21		14	15	16	17	18	19	20
	22	23	24	25	26	27	28		21	22	23	24	25	26	27
	29	30							28	29	30	31			
MAY.			1	2	3	4	5	**NOV.**					1	2	3
	6	7	8	9	10	11	12		4	5	6	7	8	9	10
	13	14	15	16	17	18	19		11	12	13	14	15	16	17
	20	21	22	23	24	25	26		18	19	20	21	22	23	24
	27	28	29	30	31				25	26	27	28	29	30	
JUNE.						1	2	**DEC.**							1
	3	4	5	6	7	8	9		2	3	4	5	6	7	8
	10	11	12	13	14	15	16		9	10	11	12	13	14	15
	17	18	19	20	21	22	23		16	17	18	19	20	21	22
	24	25	26	27	28	29	30		23	24	25	26	27	28	29
									30	31					

ECLIPSES FOR THE YEAR 1866.

There will be five Eclipses this year; three of the Sun, and two of the Moon.

I. A partial Eclipse of the Sun, March 16th. This is a small Eclipse, visible only in the extreme northeastern part of Asia and northwestern part of America.

II. A total Eclipse of the Moon, March 30th. This Eclipse is visible in this country.

III. A partial Eclipse of the Sun, April 15th. A small Eclipse, visible in Australia only.

IV. A total Eclipse of the Moon, September 24th. This Eclipse will not be visible in this country.

V. A partial Eclipse of the Sun, October 8th. This Eclipse is not visible in this country, except in the northeastern portion of it, whore it will be seen as a small partial Eclipse. It will also appear as such in the western part of Europe and the northwestern part of Africa.

EQUINOXES AND SOLSTICES.

	D.	H.	M.
Vernal Equinox	March 20,	2	46 mo.
Summer Solstice	June 21,	11	26 aft.
Autumnal Equinox	Sept. 23,	1	43 mo.
Winter Solstice	Dec. 21,	7	42 aft.

TO GAS ENGINEERS AND SUPERINTENDENTS OF GAS WORKS.

GENTLEMEN :—

Our house commenced the manufacture of gas meters eighteen years ago, and was among the first in the business in this country.

We had to contend with the prepossessions for the English made meter, and with the prejudices against those made in our own land.

Our practical and mechanical knowledge of meter making, with industry, energy, and close attention to business, soon enabled us to enter into *competition*, the result of which was a reduction in the cost of meters to gas companies. The price of a 5-light meter, which is the average selling size, at that time was $13 50. By degrees we succeeded in bringing it down to $10—our price previous to the breaking out of the war.

So far we have been successful in preventing a monopoly of the business, and have honorably endeavored to thwart every attempt tending to a consolidation of the trade.

Efforts have been made to render subservient to one combination the only meter dial maker in this country.

We think it will be hard work to persuade those interested in gas companies that these repeated efforts to monopolize the trade and crush out competition are made only for their benefit.

While the names of other firms have either passed out of the business, or changed forms, we still retain our name, and remain at this day the oldest firm in the meter trade in the United States. We think you will agree with us that the gas interest will insure a better meter and fairer prices by keeping up a healthy competition than by fostering a combination whose influence may control the trade and dictate terms.

It will hardly be necessary to say that our meters have long enjoyed the highest reputation. They have been subjected to the severest trials. Repeated tests have been made of their accuracy, durability, the workmanship, and of the materials used in their manufacture. After frequent comparisons with those of other meter makers, John C. Cresson, Esq., late Engineer of the Philadelphia Gas Works, gave us, in 1857, a certificate which will be found in our circular.

From these facts we have the confidence to bespeak a portion of your trade, and hope you will extend to us the hand of encouragement, and at least a portion of your patronage. Our long experience, aided by skilful and competent workmen, and the supervision given to the factory by one of the firm, a practical meter maker, enables us to warrant our meters and guaranty their giving entire satisfaction. If not found satisfactory upon a fair trial, we only ask their return.

Our primary object in issuing the Pocket Companion is to advertise our house, and prevent a monopoly of the business.

Our secondary object is to furnish such hints as may be useful to those engaged in the management of gas works.

Any hints from engineers or superintendents which will add to its value will be thankfully received by

DR. W. H. M'FADDEN,
OF
HARRIS & BROTHER,
No. 1117 Cherry Street, Philadelphia.

EXPLANATION OF THE GAS METER INDEXES.

DIAGRAM OF THE FACE.

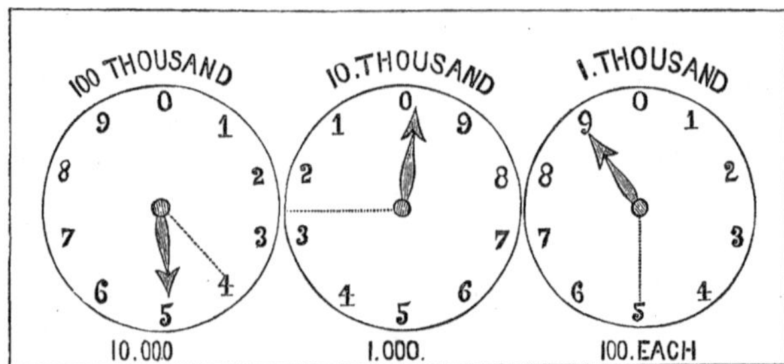

Each figure in the right hand circle denotes hundreds. When the hand completes the circle it denotes ten hundred, which is registered by the hand of the centre circle pointing to one. Each figure in the centre circle denotes one thousand. When the hand of the centre circle completes the circle it denotes ten thousand, which is registered on the left hand circle by the hand there denoting for each figure ten thousand.

The register always shows the quantity that has passed through since the meter was first set, from which deduct the amount paid for, and the difference shows the quantity remaining unpaid.

The diagram indicates	9,900
Previous observation, dotted lines	42,500
	7,400

STATION METERS.

All gas works will find it to their advantage to use a station meter. By means of it the amount of leakage or loss can be determined by a comparison of the amount made with the amount used. We make them of capacity to record from 20,000 to 1,500,000 cubic feet in 24 hours, furnished with pressure and water-line gauges, and either with a plain index or with clock and tell-tale movements complete. We would call particular attention to the improvements we have introduced, whereby the capacity is largely increased.

WET METERS.

The many certificates of Gas Engineers and Superintendents who have used our wet meters, the frequent analyses and examinations which we have made of other manufacturers' work, and the confidence which our many years' practical experience imparts, give us reason to believe our wet meters are equal, if not superior, to any made, as well in accuracy and durability as in workmanship. Many of which have been in constant use for fifteen years in the Philadelphia and other gas works, and are in good working order to-day. A sworn inspector tests and seals each meter before it leaves the factory. None but the best workmen are employed, and we use the very best material, as the certificate of N. Trotter & Co.—the largest importers of our materials in this city—will show.

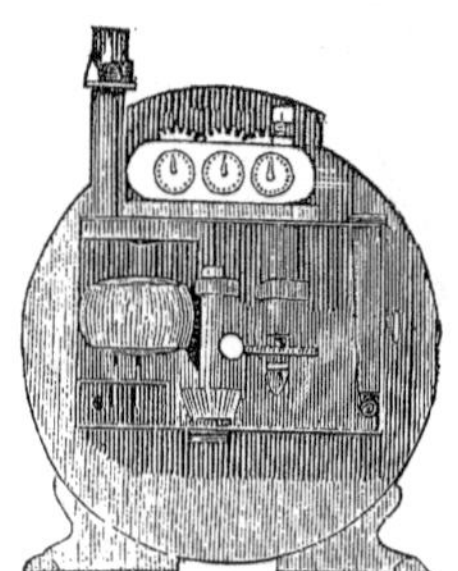

THE DRY METER.

We manufacture the two standard dry meters; the 3 diaphragm with a rotary valve, and the 2 diaphragm with a slide valve. Many advantages are justly claimed for the dry meter. It requires no fluid, as in the wet, to operate it, which is so liable to freeze during the winter months, giving constant annoyance, and subjecting consumers to frequent inconvenience and expense to furnish alcohol. It is cleaner, and

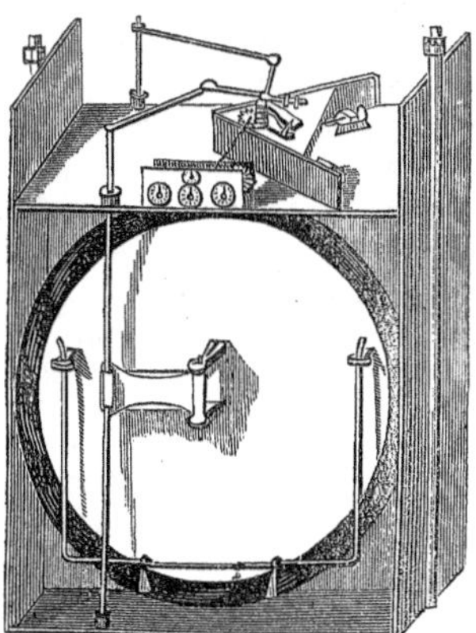

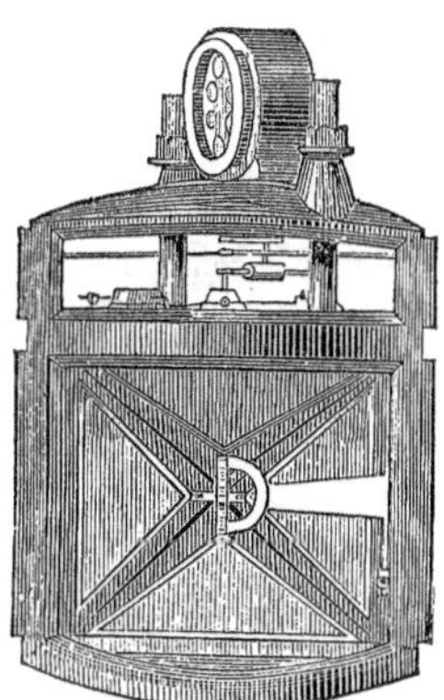

no unpleasant odor attends upon its use. It does not demand constant attention to keep up the water level, thus incurring expense to companies. Nor are consumers put to the inconvenience of light going out for want of filling up, or because too much of the fluid has evaporated. Having less occasion to open and examine it, it is not so liable to accidents nor to be tampered with by dishonest customers.

Some Engineers prefer the 3 diaphragm, yet the simplicity of the 2 diaphragm with the slide valve, and the general satisfaction it affords, have a tendency to bring it into more general use.

In our manufacture of the dry meter we give great care in selecting the cape-skins from which we make the bellows, and upon which its durability depends, and the preparation to which we subject it renders it such as cannot be surpassed for meter purposes. The small and accurate rubbing surface of the valves has enabled us, after repeated experiments, to overcome its lifting, which was heretofore such a serious objection to its accuracy. As we make them they can be certainly depended upon, and we warrant both varieties of our dry meter to give entire satisfaction.

In purchasing the dry meter, its *capacity* should receive the attention of Companies as well as the quality of the leather used in them, and the character of the rubbing surface of the valves, upon which depends the value of the meter. A good meter prevents loss to the Company, and its accuracy inspires the consumer with confidence. We feel no better security can be offered to Companies than our practical knowledge, long experience, and established reputation.

GLAZED METERS.

Consumers generally doubt the accuracy of their meters. To remove this prejudice, every company should have a glazed meter which exhibits the internal structure and working apparatus. Its operation satisfies consumers and renders their intercourse more agreeable and pleasant.

Test or Experimental Meters are of value to test the amount of gas consumed by burners. By observations of one minute you can tell how much any burner will consume in one hour. Thus testing the statements of those interested in the sale of patent burners, and quieting the complaints of consumers subjected to such cases of imposition.

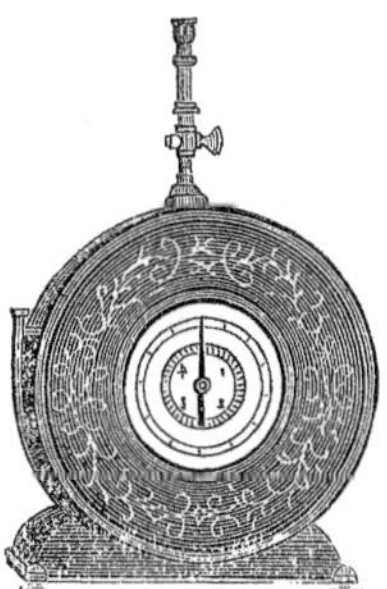

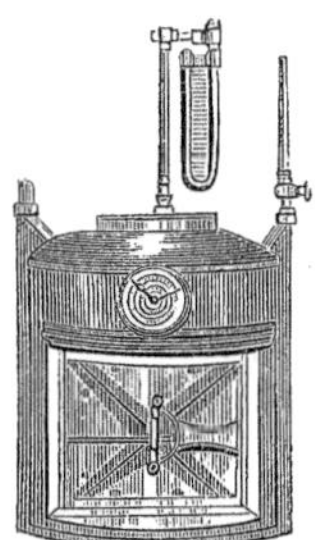

PRESSURE REGISTER.

The pressure register, which is of great value to many gas works, we get up in the neatest manner and after the most approved style. The pressure in the street mains varies with the locality and during the several hours of the day and night. This instrument gives a daily record of this variation. They are finished either plain or ornamental, and are made for various ranges.

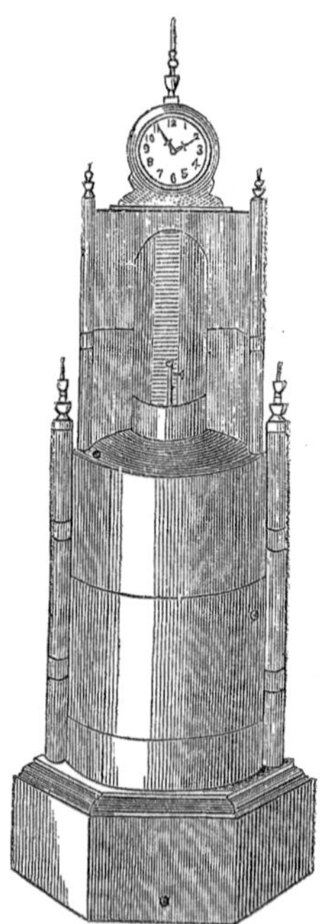

PHOTOMETER.

The photometer, invented by Count Rumford and improved by Bunsen and others, is an instrument for measuring the intensity of gas light as compared with any other standard. We keep this instrument constantly on hand, and furnish all the necessary apparatus for conducting such experiments. The chemical method of testing the intensity of light depends upon the amount of oxygen necessary to its combustion. This plan was used by Dr. Henry, and has been lately introduced into this country. This latter is called the chemical method, while the former was the optical method made by a comparison of shadows.

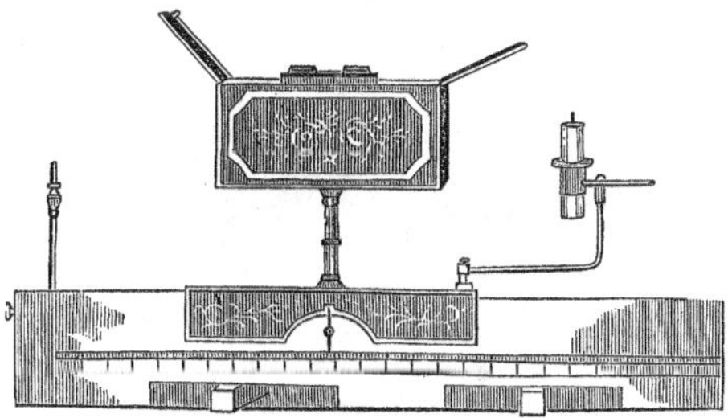

METER PROVER.

The cubic foot measures of the U. S. Assay office is the standard by which the accuracy of provers is regulated. The pressure varies with the raising and lowering of the prover holder. To overcome this and insure a uniform pressure throughout, that is, from top to bottom, we have adopted the compensating balance with its adjustment, thus securing a prover of accuracy and reliability for gas or air.

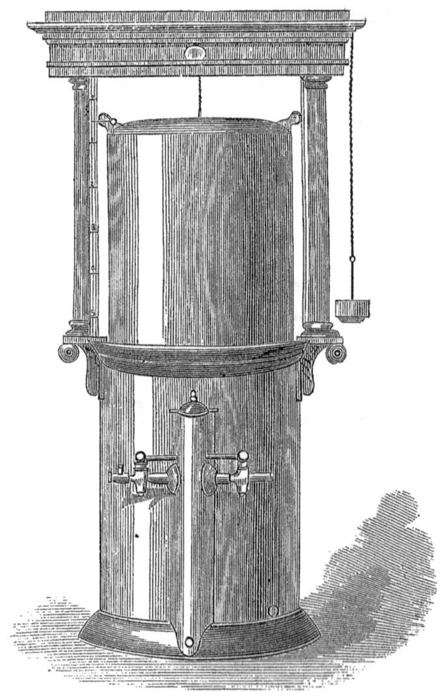

We would suggest to gas companies the propriety of having a meter prover. It would afford satisfaction not only to the Company but also to the consumers. The many disputes arising about the unfaithfulness of the meter could thus be readily met and overcome.

PRESSURE INDICATOR.

The pressure indicator is a very sensitive and reliable instrument. They are used in many works to detect leaks and as a substitute for large pressure gauges. We make two sizes, one of 6 inches the other of 9 inches, and finished either plain or ornamental.

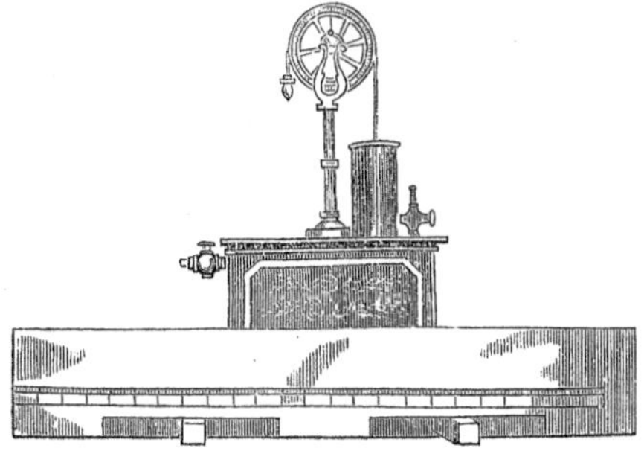

TESTING METERS.

To prove a meter is a very nice and delicate operation. To insure accuracy, certain rules are essential, and the strictest attention should be given to observe them.

1st. The prover should be mathematically correct.

2d. For a uniform pressure from top to bottom, the meter prover should be carefully counterpoised and adjusted with a compensating balance.

3d. The air in the proving room, also the air or gas, and the water in the holder, should be kept uniformly of the same temperature.

4th. A thermometor in the air or gas-holder and one in the proving room will always indicate the uniformity of the temperature or any variation which arises.

Due attention paid to these points, connect the meter to the holder. Then try your connections if they be tight. To do this you require a pressure gauge between the meter prover and meter.

Place your hand tightly upon the outlet of the meter, turn on the air or gas from the holder, then turn it off, and if the pressure gauge stands at its initial point, you are ready to go on with the proving; after having brought the hand on the index of the meter to a designated point and the holder starting from 0 or any other fixed point, observe carefully these two points, the one on the holder and the other on the index of the meter. Thus prepared, turn on to the meter the gas or air from the holder and make one or more complete revolutions of the pointer on dial, *always stopping exactly on the point started from.*

If the meter and holder exactly correspond, the meter is correct; if any variation, the percentage of error is easily calculated. To prove wet meters the only additional observations are to set them on a level plate and allow the water to run as low as it will from the water line or side screw—the true water level. Meters are usually proved under from 1 to 1½ inch pressure at inlet pipe, which should be well supplied with air or gas, and the outlet reduced to a given quantity per hour, according to the English standard as per table below.

English standard of outlet openings when tested.

2	light	Meter =	12	feet	per hour.
3	"	"	18	"	"
5	"	"	30	"	"
10	"	"	60	"	"
20	"	"	120	"	"
30	"	"	180	"	"
45	"	"	270	"	"
60	"	"	360	"	"
80	"	"	480	"	"
100	"	"	600	"	"

The lighting of the public lamps is either partial or during the absence of sunlight. The calendar will serve as a guide, and will be found very useful.

CALCULATOR OF PUBLIC LIGHTING.

Light ½ hour after sun sets.
" 1 " before moon sets.
Extinguish 1 hour before sun rises.
" 1 " after moon rises.

JANUARY, 1866.

DATE.	DAY OF WEEK.	PUBLIC LAMPS.				GAS.		
		Light.		Exting'h.		Made.	Consumed.	On hand.
		H.	M.	H.	M.			
1	M.	Full Moon.		—	—			
2	T.	5	15	6	45			
3	W.	5	15	7	45			
4	T.	5	15	8	45			
5	F.	5	15	9	45			
6	S.	5	15	10	45			
7	**S.**	5	15	11	45			
8	M.	5 L. Q.	15	Morn.				
9	T.	5	15	12	45			
10	W.	5	30	2	30			
11	T.	5	30	3	30			
12	F.	5	30	4	15			
13	S.	5	30	5	15			
14	**S.**	5	30	5	30			
15	M.	5	30	5	30			
16	T.	5 N. M.	30	5	30			
17	W.	5	30	5	30			
18	T.	5	30	5	30			
19	F.	5	30	5	30			
20	S.	5	30	5	30			
21	**S.**	9	30	5	30			
22	M.	10	45	5	30			
23	T.	11 F. Q.	45	5	30			
24	W.	Morning.		5	30			
25	T.	1	—	5	30			
26	F.	2	—	5	30			
27	S.	2	—	5	30			
28	**S.**	4	—	5	30			
29	M.	No Lighting		—	—			
30	T.	Full Moon.		—	—			
31	W.	No Lighting.		—	—			

COAL.	COKE.			MEMORANDUM.
Carbonized.	Made.	Used.	On hand.	

FEBRUARY, 1866.

DATE.	DAY OF WEEK.	PUBLIC LAMPS.				GAS.		
		Light.		Exting'h.		Made.	Consumed.	On hand.
		H.	M.	H.	M.			
1	T.	5	45	7	30			
2	F.	5	45	8	30			
3	S.	5	45	9	30			
4	**S.**	6	—	10	30			
5	M.	6	—	11	30			
6	T.	6	—	Morn.				
7	W.	6 L. Q.	—	12	15			
8	T.	6	—	1	—			
9	F.	6	—	2	—			
10	S.	6	—	2	45			
11	**S.**	6	—	3	45			
12	M.	6	—	4	45			
13	T.	6	—	5	15			
14	W.	6	—	5	15			
15	T.	6 N. M.	—	5	15			
16	F.	6	—	5	15			
17	S.	6	15	5	15			
18	**S.**	6	15	5	15			
19	M.	9	30	5	15			
20	T.	10	45	5	15			
21	W.	11 F. Q.	45	5	15			
22	T.	Morning.		5	15			
23	F.	1	—	5	15			
24	S.	2	—	5	15			
25	**S.**	2	45	5	15			
26	M.	3	45	5	15			
27	T.	3	45	5	15			
28	W.	No Lighting.		—	—			

COAL.	COKE.			MEMORANDUM.
Carbonized.	Made.	Used.	On hand.	

MARCH, 1866.

DATE.	DAY OF WEEK.	PUBLIC LAMPS.				GAS.		
		Light.		Exting'h.		Made.	Consumed.	On hand.
		H.	M.	H.	M.			
1	T.	Full Moon.		—	—			
2	F.	No Lighting.		—	—			
3	S.	6	30	8	15			
4	**S.**	6	30	9	15			
5	M.	6	30	10	15			
6	T.	6	30	11	—			
7	W.	6	30	12	—			
8	T.	6	30	Morn.				
9	F.	6 L. Q.	30	2	—			
10	S.	6	30	3	30			
11	**S.**	6	30	4	15			
12	M.	6	30	4	30			
13	T.	6	30	4	30			
14	W.	6	30	4	30			
15	T.	6	30	4	30			
16	F.	6 N. M.	45	4	30			
17	S.	6	45	4	30			
18	**S.**	6	45	4	15			
19	M.	6	45	4	15			
20	T.	6	45	4	15			
21	W.	10	45	4	15			
22	T.	11	45	4	15			
23	F.	F. Q., Morn.		4	15			
24	S.	12	45	4	15			
25	**S.**	1	30	4	15			
26	M.	2	15	4	15			
27	T.	3	—	4	15			
28	W.	3	—	4	15			
29	T.	No Lighting.		—	—			
30	F.	Full Moon.		—	—			
31	S.	No Lighting.		—	—			

COAL.	COKE.			MEMORANDUM.
Carbonized.	Made.	Used.	On hand.	

APRIL, 1866.

DATE.	DAY OF WEEK.	PUBLIC LAMPS.				GAS.		
		Light.		Exting'h.		Made.	Consumed.	On hand.
		H.	M.	H.	M.			
1	**S.**	7	—	8	—			
2	M.	7	—	9	—			
3	T.	7	—	10	—			
4	W.	7	—	10	45			
5	T.	7	—	11	45			
6	F.	7	—	Morn.				
7	S.	7	—	2	—			
8	**S.**	7 L. Q.	—	2	45			
9	M.	7	—	3	30			
10	T.	7	—	3	45			
11	W.	7	—	3	45			
12	T.	7	—	3	45			
13	F.	7	—	3	45			
14	S.	7	—	3	45			
15	**S.**	7 N. M.	15	3	45			
16	M.	7	15	3	30			
17	T.	7	15	3	30			
18	W.	7	15	3	30			
19	T.	10	45	3	30			
20	F.	11	30	3	30			
21	S.	F. Q. Morn.		3	30			
22	**S.**	12	15	3	30			
23	M.	1	—	3	30			
24	T.	1	30	3	30			
25	W.	2	—	3	30			
26	T.	2	30	3	30			
27	F.	2	30	3	30			
28	S.	No Lighting.		—	—			
29	**S.**	Full Moon.		—	—			
30	M.	No Lighting.		—	—			

COAL.	COKE.			MEMORANDUM.
Carbonized.	Made.	Used.	On hand.	

4 25

MAY, 1866.

DATE.	DAY OF WEEK.	PUBLIC LAMPS.				GAS.		
		Light.		Exting'h.		Made.	Consumed.	On hand.
		H.	M.	H.	M.			
1	T.	7	30	8	45			
2	W.	7	30	9	45			
3	T.	7	30	10	30			
4	F.	7	30	11	15			
5	S.	7	30	12	—			
6	**S.**	7	30	Morn.				
7	M.	7 L. Q.	30	12	45			
8	T.	7	30	1	15			
9	W.	7	30	2	—			
10	T.	7	30	2	30			
11	F.	7	30	3	15			
12	S.	7	30	3	15			
13	**S.**	7	30	3	15			
14	M.	7 N. M.	45	3	15			
15	T.	7	45	3	15			
16	W.	7	45	3	15			
17	T.	7	45	3	15			
18	F.	10	15	3	15			
19	S.	11	—	3	15			
20	**S.**	11	30	3	15			
21	M.	F. Q., Morn.		3	15			
22	T.	12	15	3	15			
23	W.	12	45	3	15			
24	T.	1	15	3	—			
25	F.	1	45	3	—			
26	S.	2	—	3	—			
27	**S.**	2	—	3	—			
28	M.	No Lighting.		—	—			
29	T.	Full Moon.		—	—			
30	W.	No Lighting.		—	—			
31	T.	7	45	9	15			

COAL.	COKE.			MEMORANDUM.
Carbonized.	Made.	Used.	On hand.	

JUNE, 1866.

DATE.	DAY OF WEEK.	PUBLIC LAMPS.				GAS.		
		Light.		Exting'h.		Made.	Consumed.	On hand.
		H.	M.	H.	M.			
1	F.	7	45	10	—			
2	S.	7	45	10	45			
3	**S.**	8	—	11	30			
4	M.	8	—	Morn.				
5	T.	8	—	12	15			
6	W.	8 L. Q.	—	12	45			
7	T.	8	—	2	—			
8	F.	8	—	2	30			
9	S.	8	—	3	—			
10	**S.**	8	—	3	—			
11	M.	8	—	3	—			
12	T.	8 N. M.	—	3	—			
13	W.	8	—	3	—			
14	T.	8	—	3	—			
15	F.	8	—	3	—			
16	S.	8	—	3	—			
17	**S.**	10	15	3	—			
18	M.	10	45	3	—			
19	T.	11 F. Q.	15	3	—			
20	W.	11	45	3	—			
21	T.	Morning.		3	—			
22	F.	12	15	3	—			
23	S.	12	45	3	—			
24	**S.**	1	15	3	—			
25	M.	2	—	3	—			
26	T.	No Lighting.		—	—			
27	W.	Full Moon.		—	—			
28	T.	No Lighting.		—	—			
29	F.	8	—	9	—			
30	S.	8	—	9	30			

COAL.	COKE.			MEMORANDUM.
Carbonized.	Made.	Used.	On hand.	

JULY, 1866.

DATE.	DAY OF WEEK.	PUBLIC LAMPS.				GAS.		
		Light.		Exting'h.		Made.	Consumed.	On hand.
		H.	M.	H.	M.			
1	S.	8	—	10	15			
2	M.	8	—	10	45			
3	T.	8	—	11	30			
4	W.	8	—	12	—			
5	T.	8 L. Q.	—	Morn.				
6	F.	8	—	12	30			
7	S.	8	—	1	45			
8	S.	8	—	2	30			
9	M.	8	—	3	15			
10	T.	8	—	3	15			
11	W.	8	—	3	15			
12	T.	8 N. M.	—	3	15			
13	F.	8	—	3	15			
14	S.	8	—	3	15			
15	S.	8	—	3	15			
16	M.	8	—	3	15			
17	T.	9	45	3	15			
18	W.	10	15	3	15			
19	T.	10 F. Q.	45	3	15			
20	F.	11	15	3	15			
21	S.	12	—	3	15			
22	S.	Morning.		3	15			
23	M.	12	30	3	15			
24	T.	1	15	3	15			
25	W.	2	—	3	15			
26	T.	No Lighting.		—	—			
27	F.	Full Moon.		—	—			
28	S.	No Lighting.		—	—			
29	S.	7	45	8	45			
30	M.	7	45	9	30			
31	T.	7	45	10	—			

COAL.	COKE.			MEMORANDUM.
Carbonized.	Made.	Used.	On hand.	

AUGUST, 1866.

DATE.	DAY OF WEEK.	PUBLIC LAMPS.				GAS.		
		Light.		Exting'h.		Made.	Consumed.	On hand.
		H.	M.	H.	M.			
1	W.	7	45	10	45			
2	T.	7	45	11	15			
3	F.	7 L. Q.	45	11	45			
4	S.	7	45	Morn.				
5	**S.**	7	45	12	30			
6	M.	7	45	2	15			
7	T.	7	30	3	—			
8	W.	7	30	3	45			
9	T.	7	30	3	45			
10	F.	7 N. M.	30	3	45			
11	S.	7	30	3	45			
12	**S.**	7	30	3	45			
13	M.	7	30	3	45			
14	T.	7	30	3	45			
15	W.	7	30	3	45			
16	T.	9	15	3	45			
17	F.	9	45	3	45			
18	S.	10 F. Q.	30	3	45			
19	**S.**	11	15	3	45			
20	M.	12	—	3	45			
21	T.	Morning.		3	45			
22	W.	12	45	3	45			
23	T.	1	45	3	45			
24	F.	No Lighting.		—	—			
25	S.	Full Moon.		—	—			
26	**S.**	No Lighting.		—	—			
27	M.	7	15	8	30			
28	T.	7	15	8	45			
29	W.	7	15	9	15			
30	T.	7	15	10	—			
31	F.	7	15	10	30			

COAL.	COKE.			MEMORANDUM.
Carbonized.	Made.	Used.	On hand.	

SEPTEMBER, 1866.

DATE.	DAY OF WEEK.	PUBLIC LAMPS.				GAS.		
		Light.		Exting'h.		Made.	Consumed.	On hand.
		H.	M.	H.	M.			
1	S.	7 L. Q.	—	11	15			
2	**S.**	7	—	12	—			
3	M.	7	—	Morn.				
4	T.	7	—	1	30			
5	W.	7	—	2	30			
6	T.	7	—	4	—			
7	F.	7	—	4	—			
8	S.	6 N. M.	45	4	—			
9	**S.**	6	45	4	—			
10	M.	6	45	4	—			
11	T.	6	45	4	—			
12	W.	6	45	4	—			
13	T.	6	45	4	15			
14	F.	8	30	4	15			
15	S.	9	—	4	15			
16	**S.**	9 F. Q.	45	4	15			
17	M.	10	45	4	15			
18	T.	11	30	4	15			
19	W.	Morning.		4	15			
20	T.	12	30	4	15			
21	F.	1	15	4	15			
22	S.	2	30	4	15			
23	**S.**	No Lighting.		—	—			
24	M.	Full Moon.		—	—			
25	T.	No Lighting.		—	—			
26	W.	6	15	7	45			
27	T.	6	15	8	30			
28	F.	6	15	9	15			
29	S.	6	15	10	—			
30	**S.**	6	15	11	—			

COAL.	COKE.			MEMORANDUM.
Carbonized.	Made.	Used.	On hand.	

OCTOBER, 1866.

DATE.	DAY OF WEEK.	PUBLIC LAMPS.				GAS.		
		Light.		Exting'h.		Made.	Consumed.	On hand.
		H.	M.	H.	M.			
1	M.	6 L. Q.	15	11	45			
2	T.	6	15	Morn.				
3	W.	6	15	12	45			
4	T.	6	15	1	45			
5	F.	6	—	2	45			
6	S.	6	—	3	45			
7	**S.**	6	—	4	45			
8	M.	6 N. M.	—	4	45			
9	T.	6	—	4	45			
10	W.	6	—	4	45			
11	T.	6	—	4	45			
12	F.	6	—	4	45			
13	S.	6	—	4	45			
14	**S.**	8	30	4	45			
15	M.	9	15	4	45			
16	T.	10 F. Q.	15	4	45			
17	W.	11	—	4	45			
18	T.	12	—	4	45			
19	F.	Morning.		5	—			
20	S.	1	—	5	—			
21	**S.**	2	15	5	—			
22	M.	No Lighting.		—	—			
23	T.	Full Moon.		—	—			
24	W.	No Lighting.		—	—			
25	T.	5	30	7	—			
26	F.	5	30	7	45			
27	S.	5	30	8	45			
28	**S.**	5	30	9	45			
29	M.	5	30	10	45			
30	T.	5 L. Q.	30	11	45			
31	W.	5	30	Morn.				

COAL.	COKE.			MEMORANDUM.
Carbonized.	Made.	Used.	On hand.	

NOVEMBER, 1866.

DATE.	DAY OF WEEK.	PUBLIC LAMPS.				GAS.		
		Light.		Exting'h.		Made.	Consumed.	On hand.
		H.	M.	H.	M.			
1	T.	5	30	12	45			
2	F.	5	30	1	45			
3	S.	5	30	2	45			
4	**S.**	5	30	3	45			
5	M.	5	30	4	45			
6	T.	5	30	5	15			
7	W.	5 N. M.	15	5	15			
8	T.	5	15	5	15			
9	F.	5	15	5	15			
10	S.	5	15	5	15			
11	**S.**	5	15	5	15			
12	M.	5	15	5	15			
13	T.	9	—	5	15			
14	W.	10	—	5	15			
15	T.	10 F. Q.	45	5	15			
16	F.	11	45	5	15			
17	S.	Morning.		5	15			
18	**S.**	1	—	5	15			
19	M.	2	—	5	15			
20	T.	3	15	5	15			
21	W.	No Lighting.		—	—			
22	T.	Full Moon.		—	—			
23	F.	No Lighting.		—	—			
24	S.	5	—	7	30			
25	**S.**	5	—	8	30			
26	M.	5	—	9	30			
27	T.	5	—	10	30			
28	W.	5 L. Q.	—	11	30			
29	T.	5	—	Morn.				
30	F.	5	—	12	30			

COAL.	COKE.			MEMORANDUM.
Carbonized.	Made.	Used.	On hand.	

DECEMBER, 1866.

DATE.	DAY OF WEEK.	PUBLIC LAMPS.				GAS.		
		Light.		Exting'h.		Made.	Consumed.	On hand.
		H.	M.	H.	M.			
1	S.	5	—	2	30			
2	**S.**	5	—	3	30			
3	M.	5	—	4	30			
4	T.	5	—	5	30			
5	W.	5	—	5	30			
6	T.	5	—	5	30			
7	F.	5 N. M.	—	5	30			
8	S.	5	—	5	30			
9	**S.**	5	—	5	30			
10	M.	5	—	5	30			
11	T.	5	—	5	30			
12	W.	5	—	5	30			
13	T.	9	45	5	30			
14	F.	10 F. Q.	45	5	30			
15	S.	11	45	5	30			
16	**S.**	Morning.		5	30			
17	M.	12	45	5	30			
18	T.	2	—	5	30			
19	W.	3	15	5	30			
20	T.	No Lighting.		—	—			
21	F.	Full Moon.		—	—			
22	S.	No Lighting.		—	—			
23	**S.**	5	15	7	--			
24	M.	5	15	8	15			
25	T.	5	15	9	15			
26	W.	5	15	10	15			
27	T.	5	15	11	30			
28	F.	5 L. Q.	15	Morn.				
29	S.	5	15	12	30			
30	**S.**	5	15	2	—			
31	M.	5	15	3	15			

COAL.	COKE.			MEMORANDUM.
Carbonized.	Made.	Used.	On hand.	

A BRIEF HISTORY OF GAS.

The tables, formulæ, and calculations in the *Companion* have been carefully examined and compiled from such authors as Haswell, Clegg, Hughes, and others, to which authorities we refer you for their accuracy.

Inflammable gas, a natural product of decomposition, was known to the ancients. Its source pointed out, in 1667, by Mr. Shirley, who traced it to the coal-beds under the burning Spring of Wigan, in Lancashire, England, on the surface of which the gas collected. He suggested the possibility of procuring gas from the combustion of coal.

Dr. Hale's experiments, published in 1726, demonstrated the making of gas from the distillation of coal. Dr. Clayton's experiments, proving the same, were, however, of an earlier date.

Mr. Spedding, in 1765, proposed to light the streets of White Haven with gas conveyed through pipes. His proposition was rejected, though he proved its practicability by lighting his own office.

Dr. Watson made known, in 1767, that gas retains its illuminating power after passing through water. This fact gave us the water-joint, which facilitates all the manipulations of gas works.

Mr. Wm. Murdoch first applied artificial made gas to purposes of lighting. In 1792, he lit up his house and office with it. He experimented upon the gas from various substances, different coals, gas-burners; knew the necessity of purifying the gas, passed it through water, but did not use lime. He lit up, in 1798, part of the Soho factory, and gave, probably, the first public exhibition of gas illumination in 1802, in honor of the peace of Amiens.

Dr. Henry and Mr. Clegg introduced the use of lime to purify the gas.

The erection of private gas works was commenced in 1805, by Mr. Murdoch and Mr. Clegg. The latter introduced a separate lime purifier. To him we are indebted for wet lime purifiers, hydraulic main, its dip-pipes, mode of attaching mouth-pieces, the governor, and the gas-meter in its earliest and most novel form. The formation of chartered companies began in 1809, and has since extended over the civilized world.

All attempts to profitably make gas from oil, resin, wood, &c., seem to have failed. Nature has furnished in coal the elements of gas, in the form most suitable, whence to generate it for public use, and afford sufficient profit to make the manufacture of it a business, now, one of prime necessity.

The products from the distillation of coal are thus classified :—

1st. The light-giving elements, as olefiant gas and the hydrocarburets.
2d. The heating elements, as hydrogen and carbonic oxide.
3d. The injurious elements, as carbonic acid, ammonia, sulphuretted hydrogen, &c.

It is of the first importance to remove the injurious elements from gas. To effect this, the process of purification is essential, and consists of the following steps :—

1st. Washing with water to remove the ammonia.
2d. A scrubber or breeze condenser which does the same, and hence is often dispensed with.
3d. Air or water condenser to cool the gas and deposit other impurities, as tar.
4th. Wet or dry lime purifiers to remove the carbonic acid and sulphuretted hydrogen gases. Soda adds to the efficiency of the lime.

COALS.

For information on this subject consult R. C. Taylor's "Statistics of Coal." In choosing gas coals avoid sulphur. The best for coke has a splintery fracture and a bright lustre. This coke gives out a great heat in furnaces. The varieties possessing the most resinous lustre, and fracture compact at right angles to the lamellæ, contain most bitumen, and are best for gas.

RETORTS.

Which are the better, iron or clay retorts, is still an unsettled question. Clay are at present in fashion, and being cheaper are likely to continue so. An exhauster seems essential to the efficient and successful working of clay; hence their absence from small works. Their modes of setting, shapes, and sizes vary with the views and tastes of engineers. Each one's own experience gives the best results and the greatest satisfaction.

Too low a heat diminishes the yield of gas and increases the tar; a too long continuation of the distillatory process lessens its illuminating power. Both these should be especially avoided in the manufacture of gas. The average yield of coke per ton of 2000 lbs. is from 30 to 32 bushels, weighing from 1000 to 1300 lbs., of which from one-half to two-thirds is used under the retorts.

The average charge for a retort is from 120 to 175 lbs. Charges are made every 4, 6, or 8 hours. The usual charge is a 4 hour one. A bench should be fired up gradually. An irregular heat is very destructive upon retorts.

For iron a cherry red is a good heat. For clay an orange red or white heat. Small gas works charge with a shovel; large works use a scoop.

Before drawing a charge, to prevent accidents from an explosive mixture of the gas with air, the lids should be loosened and the escaping gas lighted.

Spent lime, with clay mixed into a mortar, is used for luting the lid.

Four feet of gas for each pound of coal is the standard yield. The extremes of yield are from 3½ to 5 feet. To prevent the collection of tar and ammonia, which impedes the action of the lime, the gas should be thoroughly cooled, and condensed before reaching the purifiers, otherwise vapors pass which corrode the cocks, fittings, &c.

The tar produced per ton is 100 to 140 lbs. Of ammoniacal liquor 10 to 13 gallons. One bushel quicklime makes two slaked, which will purify 5000 feet of gas, and covers 25 square feet 2½ inches thick. Test every day each purifier with a piece of white paper dipped into water, in which sugar of lead is dissolved. The nitrate of silver is a more delicate test, but requires to be kept in an earthen bottle, or one lined with tin foil, to exclude the light. An elevation of 100 feet increases the pressure 1 inch; a depression of 100 feet diminishes it 1 inch.

A good superintendent always saves his salary: that is, the right man in the right place is economy for any company. The leakage of a gas works varies from 10 to 50 per cent.; overcoming this would save much to companies. A leakage of 10 per cent. is allowable, including condensation, &c. &c.

Mains leak either from imperfect pipe or faulty joints. Pipes are generally proved by the manufacturer. A perfect pipe rings when struck; a cracked one jars. They should be tested when laid, by an air-pump and pressure gauge, which, allowing the mercury to fall, denotes a leak. At the same time joints might be tested with soap-suds—a bubble forming where there is a leak. If tested with gas, a light would detect the leak. The former is a more delicate test, and one generally used by gas-fitters.

Where a ½ mile of main, of 3 inches diameter, deposits in the drips more than ½ a gallon in the year, a leak may be suspected, to correct which is economy.

HINTS ON SERVICES, METERS, &c.

This table is the standard of the Philadelphia Gas Works. It governs the size of pipe used by gas-fitters for consumers, and will be found of value. From it small works can determine the size of their mains, which should never be under size, as the difference in cost is not proportionate to the advantages.

Size of Tubing.	Greatest length allowed.	Greatest No. of Burners.
$\frac{1}{4}$ inch	6 feet	1 burner
$\frac{3}{8}$ "	20 "	3 burners
$\frac{1}{2}$ "	30 "	6 "
$\frac{5}{8}$ "	40 "	12 "
$\frac{3}{4}$ "	50 "	20 "
1 "	70 "	35 "
$1\frac{1}{4}$ "	100 "	60 "
$1\frac{1}{2}$ "	150 "	100 "
2 "	200 "	200 "

Size of Meters.	Greatest No. of Burners.
3 light	5 burners
5 "	10 "
10 "	20 "
20 "	40 "
30 "	60 "
45 "	100 "
100 "	250 "

An expression of the relative cost of illuminating agents:—

Gas equal to	1
Sperm oil burnt in argand	8
Mould tallow candles 6 to the lb. . .	12
Sperm oil burnt in an open lamp . .	17
Sperm candles 6 to the lb. . . .	24
Composition " "	29
Wax " "	30

TABLE

OF

CIRCUMFERENCES AND AREAS OF CIRCLES.

In Feet and 10ths of a Foot, or in Inches and 10ths of an Inch.

(From Haswell, p. 91.)

Diam.	Circum.	Area.	Diam.	Circum.	Area.	Diam.	Circum.	Area.
·0	·000	·000	3 ·0	9·424	7·068	6 ·0	18·849	28·274
·1	·314	·007	·1	9·737	7·547	·1	19·163	29·224
·2	·628	·031	·2	10·051	8·042	·2	19·477	30·190
·3	·942	·070	·3	10·366	8·553	·3	19·792	31·172
·4	1·256	·125	·4	10·680	9·079	·4	20·106	32·169
·5	1·571	·196	·5	10·994	9·621	·5	20·420	33·183
·6	1·884	·282	·6	11·308	10·178	·6	20·734	34·212
·7	2·199	·384	·7	11·622	10·752	·7	21·048	35·256
·8	2·513	·502	·8	11·936	11·341	·8	21·362	36·316
·9	2·827	·636	·9	12·251	11·945	·9	21·677	37·412
1 ·0	3·141	·785	4 ·0	12·566	12·566	7 ·0	21·991	38·484
·1	3·456	·950	·1	12·880	13·202	·1	22·305	39·592
·2	3·769	1·130	·2	13·194	13·854	·2	22·619	40·715
·3	4·084	1·327	·3	13·508	14·522	·3	22·933	41·853
·4	4·399	1·539	·4	13·822	15·205	·4	23·247	43·008
·5	4·713	1·767	·5	14·137	15·904	·5	23·562	44·178
·6	5·027	2·010	·6	14·451	16·619	·6	23·876	45·364
·7	5·341	2·269	·7	14·765	17·349	·7	24·190	46·566
·8	5·655	2·544	·8	15·079	18·095	·8	24·504	47·783
·9	5·969	2·835	·9	15·393	18·857	·9	24·818	49·016
2 ·0	6·283	3·141	5 ·0	15·708	19·635	8 ·0	25·132	50·265
·1	6·596	3·463	·1	16·022	20·428	·1	25·446	51·530
·2	6·910	3·801	·2	16·336	21·237	·2	25·760	52·810
·3	7·224	4·154	·3	16·650	22·061	·3	26·074	54·106
·4	7·538	4·523	·4	16·964	22·902	·4	26·388	55·417
·5	7·853	4·908	·5	17·278	23·758	·5	26·702	56·745
·6	8·167	5·309	·6	17·592	24·630	·6	27·016	58·082
·7	8·481	5·725	·7	17·906	25·517	·7	27·331	59·446
·8	8·795	6·157	·8	18·221	26·420	·8	27·645	60·821
·9	9·109	6·605	·9	18·535	27·339	·9	27·959	62·212

Diam.	Circum.	Area.	Diam.	Circum.	Area.	Diam.	Circum.	Area.
9 ·0	28·274	63·617	14·0	43·982	153·938	19·0	59·690	283·529
·1	28·587	65·038	·1	44·296	156·145	·1	60·004	286·521
·2	28·901	66·476	·2	44·610	158·368	·2	60·318	289·529
·3	29·215	67·929	·3	44·924	160·006	·3	60·632	292·553
·4	29·530	69·397	·4	45·238	162·860	·4	60·946	295·593
·5	29·844	70·882	·5	45·552	165·130	·5	61·260	298·698
·6	30·158	72·387	·6	45·866	167·415	·6	61·574	301·719
·7	30·472	73·898	·7	46·180	169·717	·7	61·888	304·805
·8	30·786	75·429	·8	46·495	172·034	·8	62·203	307·908
·9	31·100	76·977	·9	46·809	174·366	·9	62·517	311·026
10·0	31·416	78·540	15·0	47·124	176·715	20·0	62·832	314·160
·1	31·730	80·118	·1	47·438	179·079	·1	63·145	317·309
·2	32·044	81·713	·2	47.752	181·358	·2	63·459	320·474
·3	32·358	83·323	·3	48·066	183·854	·3	63·773	323·655
·4	32·672	84·948	·4	48·380	186·265	·4	64·088	326·852
·5	32·986	86·590	·5	48·694	188·692	·5	64·402	330·064
·6	33·300	88·248	·6	49·008	191·134	·6	64·716	333·292
·7	33·614	89·920	·7	49·323	193·593	·7	65·030	336·536
·8	33·929	91·609	·8	49·637	196·067	·8	65·344	339·795
·9	34·243	93·313	·9	49·951	198·556	·9	65·658	343·070
11·0	34·557	95.033	16·0	50·265	201·062	21·0	65·793	346·361
·1	34·871	96.769	·1	50·579	203·583	·1	66·287	349·667
·2	35·185	98.520	·2	50·893	206·120	·2	66·601	352·990
·3	35·499	100.287	·3	51·207	208·672	·3	66·915	356·328
·4	35·814	102.070	:4	51·522	211·241	·4	67·229	359·681
·5	36·128	103.860	·5	51·836	213·825	·5	67·543	363·051
·6	36·442	105.683	·6	52·150	216·424	·6	67·857	366·435
·7	36·756	107.513	·7	52·464	219·040	·7	68·171	369·837
·8	37·070	109.359	·8	52·778	221·671	·8	68·486	373·253
·9	37·384	111.220	·9	53·092	224·318	·9	68·800	376·685
12·0	37·699	113·097	17·0	53·407	226·980	22·0	69.115	380·133
·1	38·013	114·990	·1	53·721	229·658	·1	69.429	383·597
·2	38·327	116·898	·2	54·035	232·352	·2	69.743	387·076
·3	38·641	118·823	·3	54·349	235·062	·3	70.057	390·571
·4	38·955	120·763	·4	54·663	237·787	·4	70.371	394·082
·5	39·269	122·718	·5	54·977	240·528	·5	70.686	397·608
·6	39·583	124.690	·6	55·291	243·285	·6	71.000	401·150
·7	39·898	126·677	·7	55·606	246·057	·7	71.314	404·708
·8	40·212	128·679	·8	55·920	248·846	·8	71.629	408·282
·9	40·526	130·698	·9	56·234	251·650	·9	71.943	411·871
13·0	40·840	132·732	18·0	56·548	254·469	23·0	72·256	415·476
·1	41·154	134·782	·1	56·862	257·324	·1	72·570	419·097
·2	41·468	136·848	·2	57·176	260·155	·2	72·885	422·733
·3	41·782	138·929	·3	57·490	263·022	·3	73·199	426·385
·4	42·097	141·026	·4	57·805	265·905	·4	73·513	430·053
·5	42·411	143·139	·5	58·119	268·803	·5	73·827	433·737
·6	42·725	145·267	·6	58·433	271·716	·6	74·141	437·436
·7	43·039	147·411	·7	58·747	274·646	·7	74·455	441·151
·8	43·353	149·571	·8	59·061	277·591	·8	74·770	444·881
·9	43·667	151·747	·9	59·375	280·552	·9	75·084	448·628

Diam.	Circum.	Area.	Diam.	Circum.	Area.	Diam.	Circum.	Area.
24·0	75·398	452·390	27·0	84·823	572·556	30·0	94.248	706·860
·1	75·712	456·168	·1	85·137	576·805	·1	94.561	711·580
·2	76·026	459·961	·2	85·451	581·070	·2	94.875	716·316
·3	76·340	463·770	·3	85·765	585·350	·3	95.189	721·067
·4	76·655	467·595	·4	86·079	589·646	·4	95.504	725·835
·5	76·969	471·436	·5	86·393	593·958	·5	95.818	730·618
·6	77·284	475·292	·6	86·707	598·286	·6	96.132	735·417
·7	77·598	479·164	·7	87·021	602·729	·7	96.446	740·231
·8	77·912	483·052	·8	87 335	606·988	·8	96.760	745·061
·9	78·226	486·955	·9	87·649	611·363	·9	97·074	749·907
25·0	78·540	490·875	28·0	87·964	615·753	31·0	97.389	754·769
·1	78·855	494·809	·1	88·277	620·159	·1	97.703	759·646
·2	79·169	498·760	·2	88·591	624·581	·2	98.017	764·539
·3	79·483	502·726	·3	88·905	629·019	·3	98.331	769·448
·4	79·797	506·708	·4	89·220	633·472	·4	98.645	774·372
·5	80·111	510·706	·5	89·534	637·941	·5	98.959	779·313
·6	80·425	514·719	·6	89·848	642·425	·6	99.273	784·269
·7	80·739	518·748	·7	90·162	646·926	·7	99.588	789·240
·8	81·053	522·793	·8	90·476	651·442	·8	99.902	794·227
·9	81·367	526·854	·9	90·790	655·973	·9	100.216	799·230
6·0	81·681	530·930	29·0	91·106	660·521	32·0	100·531	804·249
·1	81·995	535·022	·1	91·420	665·084	·1	100·844	809·284
·2	82·309	539·129	·2	91·734	669·663	·2	101·158	814·334
·3	82·623	543·253	·3	92·048	674·258	·3	101·472	819·399
·4	82·937	547·392	·4	92·362	678·867	·4	101·787	824·481
·5	83·257	551·547	·5	92·676	683·494	·5	102·101	829·578
·6	83·556	555·717	·6	92·990	688·136	·6	102·415	834·691
·7	83 880	559·903	·7	93·305	692·793	·7	102·729	839·820
·8	84·294	564·105	·8	93·619	697·466	·8	103·043	844·964
·9	84·508	568·323	·9	93·933	702·155	·9	103·357	850·124

TABLE

OF

SQUARES, CUBES, SQUARE AND CUBE ROOTS OF NUMBERS.

(From Haswell, p. 99.)

No.	Squares.	Cubes.	Square Roots.	Cube Roots.
1	1	1	1.0000000	1.0000000
2	4	8	1·4142136	1·2599210
3	9	27	1·7320508	1·4422496
4	16	64	2·0000000	1·5874011
5	25	125	2·2360680	1·7099759
6	36	216	2·4494897	1·8171206
7	49	343	2·6457513	1·9129312
8	64	512	2·8284271	2·0000000
9	81	729	3·0000000	2 0800837
10	100	1000	3·1622777	2·1544347
11	121	1331	3·3166248	2·2239801
12	144	1728	3·4641016	2·2894286
13	169	2197	3·6055513	2·3513347
14	196	2744	3·7416574	2·4101422
15	225	3375	3·8729833	2·4662121
16	256	4096	4·0000000	2·5198421
17	289	4913	4·1231056	2·5712816
18	324	5832	4·2426407	2 6207414
19	361	6859	4·3588989	2·6684016
20	400	8000	4·4721360	2·7144177
21	441	9261	4·5825757	2 7589243
22	484	10648	4·6904158	2·8020393
23	529	12167	4·7958315	2·8438670
24	576	13824	4·8989795	2·8844991
25	625	15625	5·0000000	2·9240177
26	676	17576	5·0990195	2·9624960
27	729	19683	5·1961524	3·0000000
28	784	21952	5·2915026	3·0365839
29	841	24389	5·3851648	3·0723168
30	900	27000	5·4772256	3·1072325
31	961	29791	5·5677644	3·1413806
32	1024	32768	5·6568542	3·1748021
33	1089	35937	5·7445626	3·2075343
34	1156	39304	5·8309519	3·2396118
35	1225	42875	5·9160798	3·2710663
36	1296	46656	6·0000000	3·3019272
37	1369	50653	6·0827625	3·3322218
38	1444	54872	6·1644140	3·3619754
39	1521	59319	6·2449980	3·3912114
40	1600	64000	6·3245553	3·4199519
41	1681	68921	6·4031242	3·4482172
42	1764	74088	6·4807407	3·4760266
43	1849	79507	6·5574385	3·5033981
44	1936	85184	6·6332496	3·5303483

No.	Squares.	Cubes.	Square Roots.	Cube Roots.
45	2025	91125	6·7082039	3·5568933
46	2116	97336	6·7823300	3·5830479
47	2209	103823	6·8556546	3·6088261
48	2304	110592	6·9282032	3·6342411
49	2401	117649	7·0000000	3·6593057
50	2500	125000	7·0710678	3·6840314
51	2601	132651	7·1414284	3·7084298
52	2704	140608	7·2111026	3·7325111
53	2809	148877	7·2801099	3·7562858
54	2916	157464	7·3484692	3·7797631
55	3025	166375	7·4161985	3·8029525
56	3136	175616	7·4833148	3·8258624
57	3249	185193	7·5498344	3·8485011
58	3364	195112	7·6157731	3·8708766
59	3481	205379	7·6811457	3·8929965
60	3600	216000	7·7459667	3·9148676
61	3721	226981	7·8102497	3·9304972
62	3844	238328	7·8740079	3·9578915
63	3969	250047	7·9372539	3·9790571
64	4096	262144	8·0000000	4·0000000
65	4225	274625	8·0622577	4·0207256
66	4356	287496	8·1240384	4·0412401
67	4489	300763	8·1853528	4·0615480
68	4624	314432	8·2462113	4·0816551
69	4761	328509	8·3066239	4·1015661
70	4900	343000	8·3666003	4·1212853
71	5041	357911	8·4261498	4 1408178
72	5184	373248	8·4852814	4·1601676
73	5329	389017	8·5440037	4·1793390
74	5476	405224	8·6023253	4·1983364
75	5625	421875	8·6602540	4·2171633
76	5776	438976	8·7177979	4·2358236
77	5929	456533	8·7749644	4·2543210
78	6084	474552	8·8317609	4·2726586
79	6241	493039	8·8881944	4·2908404
80	6400	512000	8·9442719	4·3088695
81	6561	531441	9·0000000	4·3267487
82	6724	551368	9·0553851	4·3444815
83	6889	571787	9·1104336	4·3620707
84	7056	592704	9·1651514	4·3795191
85	7225	614125	9·2195445	4·3968296
86	7396	636056	9·2736185	4·4140049
87	7569	658503	9·3273791	4·4310476
88	7744	681472	9·3808315	4·4470692
89	7921	704969	9·4339811	4·4647451
90	8100	729000	9·4868330	4·4814047
91	8281	753571	9·5393920	4·4979414
92	8464	778688	9·5916630	4·5143574
93	8649	804357	9·6436508	4·5306549
94	8836	830584	9·6953597	4·5468359
95	9025	857374	9·7467943	4·5629026
96	9216	884736	9·7979590	4·5788570
97	9409	912673	9·8488578	4·5947009
98	9604	941192	9·8994949	4·6104363
99	9801	970299	9·9498744	4·6260650
100	10000	1000000	10·0000000	4·6415888

CLEGG'S TABLES OF ANALYSIS OF AMERICAN BITUMINOUS COAL.

State and County.	Locality.	Designation of Coal Beds.	By whom Analyzed.	Specific Gravity.	Analysis. Carbon.	Volatile Matter.	Ashes.
KENTUCKY	Hawsville	Splint or Cannel coal	Dr. Jackson	1.250	48.40	4.80	2.80
	Caseyville	Bituminous coal	Johnson	1.392	44.49	3.82	23.69

Fat Bituminous Coals in Western Virginia.—State Reports.

	County.	Locality.	Designation of Coal Beds.	Analysis. Carbon.	Volatile Matter.	Ashes.
[Upper Coal Series.]		Clarksburg	Main seams	56.74	41.66	1.60
		"	"	49.21	45.43	5.36
		Pruntytown	"	57.60	39.00	3.40
		Morgantown	"	60.54	37.30	2.14
Lower Coals—Valley of the Ohio.	Kanawha	1, Coal Creek	Judge Summers's Bank	55.55	41.85	2.60
	"	2, Grand Creek	"	52.75	43.20	4.05
	Logan	3, Wolf Creek, Big Sandy River	Burning Spring	47.15	48.00	4.85
	Kanawha	4, Big Coal River	(Lewis's)	50.20	47.10	2.70
	"	5, Three-mile Creek	Cartrell's	45.95	50.30	3.75
	"	6, Elk River	Friend's Mines	55.90	39.90	5.20
	Logan	7, Logan Court-house	Lawson's	58.35	39.50	2.15
	"	8, Guyandotte	Traa Fork	56.50	42.00	1.50
	"	9, Big Sandy River	Pigeon Creek	55.00	41.00	4.00

Moderately Bituminous Coals in Western Virginia.

	County.	Locality.	Designation of Coal Beds.	By whom Analyzed.	Analysis. Carbon.	Volatile Matter.	Ashes.
Lower Coal Series in the Valley of the Kanawha. Formation No. XI., Rogers.		Big Sewell Mountain, W. flank	Tyree's bed	W. B. Rogers	67.84	30.08	2.08
		"	Deem's bed	"	71.73	27.13	1.14
	Fayette	Mill Creek	Paris's bank	"	71.88	26.20	1.92
		Scrabble Creek		"	63.36	29.04	7.60
		Bell Creek		"		32.16	
		Keller's Creek	Hansford's	W. B. Rogers, State Report	60.92	37.08	2.00
		Second seam	Storkton's mine	"	74.55	21.13	4.32
		Campbell's Creek	Ruffner's second seam	"	55.76	32.44	11.80
		"	Noyes's seam	"	64.16	32.24	3.60
		"	"	"	65.64	31.28	3.08
		Cox's Creek	Third seam	"	51 41	42.55	6.04
		Faure's Bank	Upper seam	"	53.20	35.04	11.76
		L. Ruffner's Bank	"	"	49.84	44 28	5.88
		Bream's Bank	Third seam	"	57.76	33.68	8.56
		Smither's Bank		"	54.52	29.76	15.76
		Hughes's Bank		"	62.32	32.88	4.80
		D. Ruffner's Bank	Upper seam	"	57.28	35.08	7.64
		Warth's Bank		"	54.00	39.76	6.24
Western Virginia. *Preston and Monongalia Counties.—Basins containing the Lower Coal Series.*	Preston	Kingswood	Fairfax's	State Reports	53.77	31.75	14.48
	"	"	Middle seam		65.32	27.77	6.91
	"	"	Forman's basin		73.68	21.00	5.32
	"	Deck Hollow, c.	Martin's		65.42	23.42	11 16
	"	Buffalo Lech run	Beatty's		62.56	29.60	7.84
	"	N. Brandonville	Morton's		65.28	30.80	3.92
	"	Cheat River, near Kingswood	Price's		60.36	25.00	14.64
	"	Big Sandy, W. side	Seaport's		66.64	27.12	6.24
	"	Kingswood	Hagan's		68.32	26.48	5.20
	"	"			67.28	29.68	3.04
	"	Big Sandy Basin	W. side Cheat		60.04	26.88	13.08
	"	Kingswood	Cresaps		64.24	30.24	5.32

Bituminous Coals in Eastern Virginia, in the Chesterfield, Powhatan, Goochland, and Henrico Basins.

County.	Locality.	Designation of Coal Beds.	By whom Analyzed.	Analysis.		
				Carbon.	Volatile Matter.	Ashes.
South side of James River 1	Stonehenge	Chesterfield		58.70	36.50	4.80
Chesterfield 2	Maidenhead	Engine shaft		63.97	32.83	3.20
" 3	Heth's Pit	"		62.35	37.65	2.80
" 4	Mill's and Reid's	Creek pit		57.80	38.60	3.60
" 5	Will's Pit			62.90	32.50	4.60
" 6		Green-hole shaft		67.83	30 17	2.00
" 7	Heth's Deep Shaft	Bottom seam		53.36	35.82	10.82
	"	Middle seam		66.50	28.40	5.10
	"	Top seam		61.68	28 80	9.52
" 8	Powhatan Pits	Finney		59.87	32.33	7.80
" 9	Winterpock Creek	Cox's mine		65.52	29.12	5.36
	Cloverhill, Appomattox R.	Slate coal	G. W. Andrews, M. D.	55.00	38.50	6.50
	"	Mean of four species	Johnson	54 83	33.04	10.13
	Richmond coal		Andrews	59.25	32.00	8.75
	Mid Lothian	Wooldridge's pit	Johnson	61.08	28.45	10.47
	"	Mean result, average size coal	"	53.01	33.25	14 74
	Creek Coal Co.	Mean of six trials	"	60.30	31.13	8.57
	Black Heath Pits	Mean of four species	"	58.79	32.57	8.64
	Tippecanoe Pits	"	"	54.62	36.01	9.37
North side of James R. 10		Randolph's	W. B. Rogers, State Report	66.15	30.50	3.35
11	Coalbrook Dale	Second seam	"	66.48	29.00	4.52
12	Anderson's Pit	First seam	"	66.78	28.30	4.92
17	Crouche's Lower Shaft	Upper seam, 110 ft. from surface	"	64.60	30.00	5.40
18	Scott's Pit		Johnson, State Report	60.86	33.70	5.44
19	Waterloo Shaft		"	55.20	26.80	18.00
20	Deep Run Pits		"	69.84	25.16	5.00
	Wills's Pit	Upper vein	T. G. Clemson	66.60	28.80	4.60
	Anderson's Pit	Bottom seam	R. C. Taylor	64.20	26.00	9.80

Fat Bituminous Coals in the State of Ohio.

County.	Locality.	Designation of Coal Beds.	By whom Analyzed.	Specific Gravity.	Analysis.		
					Carbon.	Volatile Matter.	Ashes.
Portland	Talmadge	Upson's mine	W.W.Mather	1.264	53.404	44.298	2.288
Jackson	Lick Township		"	1.283	49.882	47.327	2.221
"	Madison Township		J. L. Cassels	1.560	39.950	44.800	14.620
"	"	Cannel coal	"	1.410			
	Carr's Run		R. C. T.	1.270			
	Pomeroy		Dr. J. Percy		76.70	18.70	4.60

Fat Bituminous Coals in Pennsylvania.

County.	Locality.	Designation of Coal Beds.	By whom Analyzed.	Specific Gravity.	Analysis.		
					Carbon	Volatile Matter.	Ashes.
Venango	Shippensville	Sandy Ridge	H. D. Rogers, State Report		49.80	43.20	7.00
"	6. M. F. of Franklin		"		29.54	52.78	17.68
Beaver	Greersburg		"		30.12	36.00	33.88
Crawford	Conneaut Lake		"		59.45	38.75	1.80
Mercer	Greensville		"		57.80	40.50	1.70
	"		R. C. T.	1.25			
	Orangeville		State Report		53.45	43.75	2.80

Bituminous Coals.

State and County.	Locality.	Designation of Coal Beds.	By whom Analyzed.	Specific Gravity.	ANALYSIS. Carbon.	Volatile Matter.	Ashes.
INDIANA.							
Vermilion	Brouillet's Cr'k		D. D. Owen	1.270	52.00	39.00	9.00
Vigo	Honey Creek		"	1.240	70.00	27.50	2.50
Sullivan	Busseron	Lick Fork	"	1.240	70.00	28.00	2.00
Fountain	Wabash	Coal Creek mouth	"	1.260	60.00	25.00	15.00
Spencer	Anderson Cr'k		"	1.270	45.00		
	White River		"	1.270	56.40		
	Terre Haute		"	1.240	50.80		
	Cannelton	Cannel coal	W. R. Johnson	1.272	59.47	36.59	3 94
ILLINOIS	Rock River	Coal	Dr. D. D. Owen	1.340	45.50	44.50	10.00
	Vermilion	Danville	A. Morfit		48.50	47.20	4.30
	Western Port		Johnson	1.290		32.80	
	Ottowa		J. F. Frazer		62.60	35.50	1.90
	Rockwell		C. U. Sheppard	1.273	46.50	47.50	6.00
IOWA	Duck Creek	West bank of the Mississippi River	Dr. D. D. Owen	1.270	48.50	44.00	7.50
		Mastodon vein, 46 ft. thick	Booth & Boye		46 83	40.05	13.12
			J. R. Chilton, M. D.	1.252	50.81	34.06	15.13
MISSOURI	Côte sans dessein, Callaway County	Mammoth vein, 24 feet	"	1.250	50.78	34.20	15.02
	Osage River		W. R. Johnson	1.200	51.16	43.50	5.34
ARKANSAS	Johnson Co.	Spaldre's bluff	J. F. Frazer	1.396	62.60	28.90	8.50
MAINE		Peat	Dr. Jackson		21.00	72.00	7.00
Miscellaneous Analysis.							
Isle of Cuba	Near Havana	Asphaltum	T. G. Clemson	1.190	34.97	63.00	2.03
	Near Matanzas	Asphaltum	"				13.50
South America	Peru	Coxitambo	M. Bousingault				
	Chili	Arauco	W. R. Johnson	1.324	67.62	30.00	2.38
	Brazil		Karsten	1.289	57.90	40.50	1.60
	"		"	1.483	38.10	33.50	28.40
Madeira Island	Brown coal	Or lignite	Johnstone				20.05
Brit. America, Bitum. Coal.							
Nova Scotia	Pictou	Cunard's sample	Johnson	1.325	60.73	26.76	12.51
		Mining Association	"	1.318	56.98	29.63	13.39
Cape Breton	Sydney	Mean of two species	"	1.338	67.57	26.93	5.50

LINEAR MEASUREMENT.

12 inches = 1 foot. 3 feet = 1 yard.

1 mile = 1,760 yards = 5,280 feet = 63,360 inches.

Lineal feet multiplied by .00019 = miles.

" yards " .000568 = miles.

1. In a *right angled triangle* the sum of the squares of the two shorter sides = The square of the hypothenuse: the square of the hypothenuse less the square of one side = the square of the third side.
2. The diameter of a CIRCLE × 3.1416 = the CIRCUMFERENCE.
3. The circumference of a circle × 0.31831 = the DIAMETER.
4. Given a chord and versed sine — to find the DIAMETER of the *circle*. Divide the square of half the chord by the versed sine, and add the versed sine to the product = the diameter.
5. To find the *length of an arc* of a CIRCLE, when the chord of the whole arc and the chord of one-half of the arc are known—from eight times the chord of one-half of the arc, subtract the chord of the whole arc: one-third of the remainder will be the length of the arc nearly.
6. PERIPHERY of an *ellipse*. Multiply the square root of the sum of the squares of the axes, by 2.22.

SURFACE MEASUREMENT.

AREAS.

Product of two Linear Dimensions (proportioned to the squares of similar sides.)

144 square inches = 1 square foot.
9 " feet = 1 " yard.
Acre = 43,560 square feet = 4,480 yards = (660 feet × 66 feet.)
Square mile = 640 acres.

1. Parallelogram—(Square, rectangular or rhomboidal) = the product of the length of one side × by perpendicular height.

2. Triangle— = product of base × by one-half the perpendicular height.

3. Triangle—Area from three sides given. From the half sum of the three sides subtract each side separately; multiply the half sum and the three remainders together, and the square root of the product will be the area.

4. Trapezoid = the sum of the two parallel sides × by half the perpendicular height.

5. Circle = the square of the diameter × ·7854, or square of the circumference × ·07958.

6. Sector of a Circle = radius of the circle × by one-half the arc of the sector.

7. Segment of a Circle. — Find the area of a sector of a circle having the same arc, and deduct the triangle formed between the two radii and the chord of the arc.

SUPERFICIAL AREA OF SOLIDS.

8. Cube.
9. Parallelopipedon.
10. Prism.

} = Sum of areas of sides and bases.

11. Cylinder. = Circumference of base × height + area of bases.
12. Cone.
13. Pyramid.

} = { Circumference of base × one-half slant height + area of base.

14. Sphere. = Square of diameter × 3·1416.

SOLID MEASUREMENT.

CUBIC CONTENT.

Product of three Linear Dimensions, (proportional to cube of similar sides.)

Cubic foot	. . . . .		= 1,728	cubic inches.
" yard	. = 27	cubic feet	= 46,656	"
Barrel	. . = 4·8125	"	= 8,316	"
Bushel	. . = 1·2438	"	= 2,150	"
Gallon (wine)	. . . . .		= 231	"

1 lb. avoirdupois = 27·7015 cubic inches of water at 39·83° Fah. and barometer 30 = 16 ounces = 256 drachms = 7000 grains.

1 lb. troy = 12 ounces = 96 drachms = 288 scruples = 5760 grains.

Ton = 2,240 lbs. avoirdupois.

1 gallon of water weighs 58372·1757 grains troy = 11·34 lbs. troy.

Cylindrical	inches	× ·0004546	=	cubic feet.
"	feet	× ·02909	=	" yard.
Cubic	inches	× ·00058	=	" feet.
"	feet	× ·03704	=	" yard.
"	feet	× 7·48	=	United States gallons.
"	inches	× ·004329	=	" " "
Cylindrical	feet	× 5·874	=	" " "
"	inches	× ·0034	=	" " "

Content of Cask.—Add into one sum 39 times the square of the bung diameter, 25 times the square of the head diameter, and 26 times the product of the two diameters; then multiply the sum by the length, and the product again by $\frac{00034}{9}$ for wine gallons.

General Rule for Finding Cubic Content contained between two parallel planes.

Let A and B be areas of ends of solids, and C the area of a section parallel to, and equidistant from the ends, and L the distance between the ends.

$$\text{Solidity} = \frac{A + B + 4\,C}{6} \times L.$$

1. CUBE = side × side × side, or = area of base × by perpendicular height.

2. PARALLELOPIPEDON. } = { Area of base × by perpendicular height.
   PRISM.
   CYLINDER.

3. Cone. } = Area of bases × by $\frac{1}{3}$ the perpendicular height.
   PYRAMID.

4. FRUSTRUM OF CONE OR PYRAMID = sum of the areas of the two ends + the square root of their product × by $\frac{1}{3}$ of the perpendicular height.

5. SPHERE = cube of the diameter × 0·5236.

6. SPHERICAL SEGMENT = 3 times the square of the radius of its base + the square of its height × by the height × 0·5236.

WEIGHT OF BAR IRON.

The table gives the weight of a flat bar $\frac{1}{16}$ inch thick, one foot long, and of width in table.

Multiply the weight in table by number of sixteenths the iron is thick for weight desired.

(Prepared from Haswell, p. 246.)

ROLLED IRON.

FLAT IRON.

Size.	Weight of one foot in length.	Size.	Weight of one foot in length.
Inch.		Inch.	
$\frac{1}{16}$ × $\frac{1}{16}$	0·0132	$\frac{1}{16}$ × $\frac{9}{16}$	0·1188
" × $\frac{1}{8}$	0·0264	" × $\frac{5}{8}$	0·1320
" × $\frac{3}{16}$	0·0396	" × $\frac{11}{16}$	0·1452
" × $\frac{1}{4}$	0·0528	" × $\frac{3}{4}$	0·1584
" × $\frac{5}{16}$	0·0660	" × $\frac{13}{16}$	0·1716
" × $\frac{3}{8}$	0·0792	" × $\frac{7}{8}$	0·1848
" × $\frac{7}{16}$	0·0924	" × $\frac{15}{16}$	0·1980
" × $\frac{1}{2}$	0·1056	" × 1	0·2112

Round Iron.

(Haswell, p. 251.)

Size.	Weight of one foot in length.	Size.	Weight of one foot in length.
$\frac{1}{8}$ inch.	·041 lb.	$\frac{5}{8}$ inch.	1·043 lb.
$\frac{3}{16}$ "	·119 "	$\frac{11}{16}$ "	1·255 "
$\frac{1}{4}$ "	·165 "	$\frac{3}{4}$ "	1·493 "
$\frac{5}{16}$ "	·261 "	$\frac{13}{16}$ "	1·752 "
$\frac{3}{8}$ "	·373 "	$\frac{7}{8}$ "	2·032 "
$\frac{7}{16}$ "	·508 "	1 "	2·654 "
$\frac{1}{2}$ "	·663 "	$1\frac{1}{8}$ "	3·360 "
$\frac{9}{16}$ "	·840 "	$1\frac{1}{4}$ "	4·172 "

Birmingham Gauge for Wire, Sheet-Iron, and Steel.

(Haswell, p. 247.)

Thickness by the Gauge.	Thickness in inches.	Wt. per sq. foot in lbs. Sheet and Boiler Iron.	Thickness by the Gauge.	Thickness in inches.	Wt. per sq. foot in lbs. Sheet and Boiler Iron.
No. 0	0·340	13·7	No. 19	0·042	1·69
" 1	0·300	12·1	" 20	0·035	1·41
" 2	0·284	11·4	" 21	0·032	1·29
" 3	0·259	10·4	" 22	0·028	1·13
" 4	0·238	9·60	" 23	0·025	1·00
" 5	0·220	8·85	" 24	0·022	0·885
" 6	0·203	8·17	" 25	0·020	0·805
" 7	0·180	7·24	" 26	0·018	0·724
" 8	0·165	6·65	" 27	0·016	0·644
" 9	0·148	5·96	" 28	0·014	0·563
" 10	0·134	5·40	" 29	0·013	0·523
" 11	0·120	4·83	" 30	0·012	0·483
" 12	0·109	4·40	" 31	0·010	0·402
" 13	0·095	3·83	" 32	0·009	0·362
" 14	0·083	3·34	" 33	0·008	0·322
" 15	0·072	2·90	" 34	0·007	0·282
" 16	0·065	2·62	" 35	0·005	0·230
" 17	0·058	2·34	" 36	0·004	0·170
" 18	0·049	1·97			

STRENGTH OF MATERIALS.

FOR TENSION ON EACH SECTION OF ONE SQUARE INCH.

Safe Weight for Ordinary Materials.

Steel	25,000 lbs.	Lead	200 lbs.
Wrought Iron .	10,000 "	Brass	1,000 "
Cast " .	2,000 "	Oak Wood . .	1,800 "
Copper, cast . .	5,000 "	Yellow Pine . .	2,000 "
" rolled .	8,000 "	White Pine . .	1,500 "

FOR COMPRESSION.

Iron, wrought	8,000 lbs.
" cast	30,000 "
Oak	1,800 "
Stone, hard	3,000 "
Sandstone	1,200 "
Brick	60 "

Compression estimated for a column 6 diameters long.

If length = 12 diameters	. . .	deduct $\frac{1}{6}$
" = 24 "	. . . .	" $\frac{2}{6}$
" = 48 "	. . .	" $\frac{3}{6}$

HEMP ROPE.

Multiply the square of the circumference by 100 for the safe weight in pounds.

Circumference	1 inch	. . .	=	100 lbs.
"	$1\frac{1}{2}$ "	. . .	=	225 "
"	2 "	. . .	=	400 "
"	$2\frac{1}{2}$ "	. . .	=	625 "
"	3 "	. . .	=	900 "
"	$3\frac{1}{2}$ "	. . .	=	1,225 "
"	4 "	. . .	=	1,600 "
"	$4\frac{1}{2}$ "	. . .	=	2,025 "
"	5 "	. . .	=	2,590 "

CHAINS.

(Haswell, p. 258.)

Inch.	Weight per foot.	Safe Weight in pounds.	Inch.	Weight per foot.	Safe Weight in pounds.
$\frac{1}{8}$	0·17	250	$\frac{5}{8}$	4·0	6,250
$\frac{3}{16}$	0·38	560	$\frac{11}{16}$	4·84	7,550
$\frac{1}{4}$	0·67	1,000	$\frac{3}{4}$	5·75	9,000
$\frac{5}{16}$	1·08	1,560	$\frac{13}{16}$	6·0	10,500
$\frac{3}{8}$	1·55	2,250	$\frac{7}{8}$	7·83	12,250
$\frac{7}{16}$	2·11	3,050	$\frac{15}{16}$	9·4	14,000
$\frac{1}{2}$	2·7	4,000	1	10·7	16,000
$\frac{9}{16}$	3·42	5,050			

WEIGHT OF CAST-IRON PIPES.

(Haswell, p. 252.)

Diameter.	Length.		Weight.	Diameter.	Length.		Weight.
Inch.	Feet.	Inch.	Pounds.	Inch.	Feet.	Inch.	Pounds.
60	12	5	10,900	10	9		500
48	12		6,500	8	9		400
36	12		5,400	6	12		340
30	12		4,000	6	9		280
30	9		3,000	4	12		240
24	9		2,100	4	9		180
20	9		1,400	3	9		112
16	12	5	1,450	3	9		105
16	9		1,000	2	6		45
12	9		600	1½	6		35

WEIGHT OF LEAD AND GASKET REQUIRED FOR STREET MAINS.

EACH JOINT REQUIRES—

	Lead.	Gasket.		Lead.	Gasket.
2 in. pipe,	3·25 lbs.	0.050 lbs.	10 in. pipe,	15· lbs.	0·30 lbs.
3 "	4·75 "	0.075 "	12 "	20· "	0·35 "
4 "	6· "	0.115 "	16 "	25· "	0·45 "
6 "	9· "	0.175 "	18 "	29· "	0.52 "
8 "	12· "	0.250 "	20 "	43· "	0.60 "

SPECIFIC GRAVITY.

(Haswell, p. 145.)

Mercury	13,600
Lead	11,325
Copper	9,000
Cast Brass	8,000
Steel	7,850
Wrought-iron	7,780
Cast-iron	7,207
Tin	7,300
Marble	2,690
Common Stone	2,520
Brick	1,900 @ 2,000
Soil	1,984
Coal, anthracite	1,436 @ 1,640
" bituminous	1,270
Sand	1,520
Sea-Water	1,030
COMMON WATER	1,000
Oak, (dry)	925
Ash, "	800
Maple, "	755
Elm, "	600
Yellow Pine, "	660
White Pine, "	554
Cork	240
Carb. Acid	1·9
Air	1·25
Coal Gas	0·6
Hydrogen	0·0848

The specific gravity in table also represents the number of ounces in each substance in 1 cubic foot ÷ 16 = lbs.

1 cubic foot of Cast-Iron	=	450 lbs.
1 " " White Pine	=	34·6 "
1 " " Water	=	62·5 "
10·9 cubic feet of Air	=	1·– "
22 " " Coal Gas	=	1·– "

GROSS TON IN BULK.

Anthracite Coal, 1 cubic foot = 53 lbs.—1 ton = 42·3 cu. ft.
Bituminous " 1 " = 50 " 1 " = 44·5 "
Charcoal 1 " = 18·2 " 1 " = 123· "

70·82 lbs. Pittsburg Coal make 1 bushel of Coke.
+ Breeze and waste (= 11 per cent.)
1 cord of wood = 8 × 4 × 4 cubic feet.
1 Ton (2,240 lbs.) Pittsburg Coal = 31½ bush. coke + 3½ bush. Breeze @ 38 lbs. per. bush.

FOR BALLOONS.

Coal-Gas—specific gravity 0·625 water.

Each 13 cubic feet of air . . . . = 1 lb.
" 13 " " coal-gas . . . = ½ "
" 13 " " " . buoyancy = ½ "

For 1,000 cubic feet coal-gas, 40 lbs. buoyancy nearly.
" " hydrogen, 75 " "

TABLE OF EXPANSION BY THERMOMETER.

Solids heated from 32° to 212° (Fahrenheit) expand—

	From 1 to	For each degree.
Wrought-Iron	1·00122045	0·00000680
Cast-Iron	1·00111120	0·00000618
Hardened Steel	1·00123956	0·00000689
Copper	1·00171820	0·00000955
Lead	1·00284836	0·00001580
Brass (common cast)	1·00187821	0·00001043
Brass Wire	1·00193333	0·00001075
Tin, hammered	1·00270000	0·00001500
" cast	1·00217298	0·00001207
Marble	1.00110410	0·00000613
LIQUIDS.		
Mercury	1·00018433	0·00001025
Water	1·00046600	0·00002595
Oil (common)	1·00080000	0·00004444
Alcohol	1·00100000	0·00005555

(From FARADAY'S CHEMICAL MANIPULATIONS.)

TABLE OF AQUEOUS VAPOR CONTAINED IN 1000 CUBIC FEET OF GAS AT INDICATED TEMPERATURE.

Temp.	Volume.	Temp.	Volume.	Temp.	Volume.
DEGREES.	A. V.	DEGREES.	A. V.	DEGREES.	A. V.
40	9·33	54	15·33	68	24·06
41	9·73	55	15·86	69	24·83
42	10·13	56	16·40	70	25·66
43	10·53	57	16·93	71	26·53
44	10·93	58	17·53	72	27·40
45	11·33	59	18·10	73	28·30
46	11·73	60	18·66	74	29·23
47	12·13	61	19·23	75	30.20
48	12·53	62	19·80	76	31·20
49	12·93	63	20·50	77	32·20
50	13·33	64	21·20	78	33·23
51	13·80	65	21·90	79	34·23
52	14·26	66	22·60	80	35·33
53	14·80	67	23·30		

FLOW OF GAS THROUGH MAINS. [PER HOUR.]

With loss of $\frac{5}{10}$ of an inch pressure.

Diameter.	50 ft.	100 ft.	500 ft.	1,000 ft.
2 inch.	2,000	1,600	700	500
3 "	4,700	3,750	1,600	1,200
4 "	8,500	7,000	2,900	2,200
6 "	19,000	15,000	6,500	5,000
8 "	36,000	29,000	13,000	10,000
10 "	65,000	52,000	23,000	18,000
12 "	100,000	80,000	36,000	30,000

With loss of 1 inch pressure.

Diameter.	50 ft.	100 ft.	500 ft.	1,000 ft.
2 inch.	3,500	2,800	1,250	875
3 "	7,900	6,300	2,800	2,000
4 "	14,000	11,500	5,000	3,600
6 "	32,000	26,000	12,000	9,000
8 "	58,000	47,000	22,000	15,000
10 "	90,000	75,000	33,000	23,500
12 "	135,000	113,000	50,000	36,000

EFFECT OF BAROMETRIC PRESSURE UPON GASES.

Additional Pressure. Feet of Water.	Relation to Original Volume.	Total Pressure in feet of Water
0·	...... 1·	...... 33 + 0 = 33
1·	...... $\frac{33}{34}$	...... 33 + 1 = 34
2·	...... $\frac{33}{35}$	...... 33 + 2 = 35
3·	...... $\frac{33}{36}$	...... 33 + 3 = 36
4·	...... $\frac{33}{37}$	...... 33 + 4 = 37
5·	...... $\frac{33}{38}$	...... 33 + 5 = 38
10·	...... $\frac{33}{43}$	...... 33 + 10 = 43
20·	...... $\frac{33}{53}$	...... 33 + 20 = 53
30·	...... $\frac{33}{63}$	...... 33 + 30 = 63
33·	...... $\frac{33}{66}$	...... 33 + 33 = 66

$\frac{33}{66} = \frac{1}{2}$ volume for 1 atmosphere additional.

PHOTOMETRY.

1 Wax Candle,	4 to a pound,	burns 13 hours.		
1 Spermaceti Candle,	6 "	" 8 "		
1 Tallow "	6 "	" 6 "		40 min.

RELATIVE LIGHT FOR UNIT OF GAS.

Batswing	consuming	5	feet,	1·000
Large Patent Argand Burner .	"	6	"	2·880
Boston Argand	"	5·4	"	2·132
Single jet	"	2·2	"	1·191
Solliday's Patent	"	5·7	"	1·
Fish-tail (Union-jet) . .	"	4.5	"	1·513
Large Batswing . . .	"	11·3	"	2·03
Wax candle, 4 to lb.				0·143
Sperm " 6 "				0·111
Tallow " 6 "				0·1

TABLE.

Showing the number of candles any other light is equal to, the centre of candle being 10 *inches from centre of Photometer, and the gas (or other light) at any distance from* 5 *to* 50 *inches.*

Distance of Gas Burner from Photom'r.	Number of Candles Gas Burner is equal to.	Dis-tance.	Can-dles.	Dis-tance.	Can-dles.	Dis-tance.	Can-dles
5 in.	·25	16 in.	2·56	27½ in.	7·56	39 in.	15·21
5½ "	·30	16½ "	2·72	28 "	7·84	39½ "	15·60
6 "	·36	17 "	2·80	28½ "	8·12	40 "	16·00
6½ "	·42	17½ "	3·06	29 "	8·41	40½ "	16·40
7 "	·49	18 "	3·24	29½ "	8·70	41 "	16·81
7½ "	·56	18½ "	3·42	30 "	9·00	41½ "	17·22
8 "	·64	19 "	3 61	30½ "	9·30	42 "	17·64
8½ "	·72	19½ "	3·80	31 "	9·61	42½ "	18·06
9 "	·81	20 "	4·00	31½ "	9·92	43 "	18·49
9½ "	·90	20½ "	4·20	32 "	10·24	43½ "	18·92
10 "	1·00	21 "	4·41	32½ "	10·56	44 "	19·36
10½ "	1·10	21½ "	4·62	33 "	10·89	44½ "	19·80
11 "	1·21	22 "	4·84	33½ "	11·22	45 "	20·25
11½ "	1·32	22½ "	5·06	34 "	11·56	45½ "	20·70
12 "	1·44	23 "	5·29	34½ "	11·90	46 "	21·16
12½ "	1·56	23½ "	5·52	35 "	12·25	46½ "	21·62
13 "	1·69	24 "	5·76	35½ "	12·60	47 "	22·09
13½ "	1·82	24½ "	6·00	36 "	12·96	47½ "	22·56
14 "	1·96	25 "	6·25	36½ "	13·32	48 "	23·04
14½ "	2·10	25½ "	6·50	37 "	13·69	48½ "	23·52
15 "	2·25	26 "	6·76	37½ "	14·06	49 "	24·01
15½ "	2·40	26½ "	7·02	38 "	14·44	49½ "	24·50
....	....	27 "	7·29	38½ "	14·82	50 "	25·00

If the distance be in inches and tenths, square the number, and point off four decimals.

Example :—

```
   13·5
   13·5
  -----
    675
   405
  135
  -----
 1·8225 candles.
```

In these tests, the English sperm standard candle is to be preferred, on account of steadiness of flame and uniformity of consumption.

CIRCLES AND AREAS OF CIRCLES, BY $\frac{1}{8}$.

Diam.	Circumf.	Area.	Diam.	Circumf.	Area.
$\frac{1}{32}$	·0981	·00076	$\frac{7}{8}$	2·748	·6013
$\frac{1}{16}$	·1963	·00306	$\frac{15}{16}$	2·945	·6902
$\frac{1}{8}$	·3926	·01227	1	3·141	·7854
$\frac{3}{16}$	·5890	·02761	1 $\frac{1}{8}$	3·534	·9940
$\frac{1}{4}$	·7854	·04908	1 $\frac{1}{4}$	3·927	1·227
$\frac{5}{16}$	·9817	·07669	1 $\frac{3}{8}$	4·319	1·484
$\frac{3}{8}$	1·178	·1104	1 $\frac{1}{2}$	4·712	1·767
$\frac{7}{16}$	1·374	·1503	1 $\frac{5}{8}$	5·105	2·073
$\frac{1}{2}$	1·570	·1963	1 $\frac{3}{4}$	5·497	2·405
$\frac{9}{16}$	1·767	·2485	1 $\frac{7}{8}$	5·890	2·761
$\frac{5}{8}$	1·963	·3067	2	6·283	3·141
$\frac{11}{16}$	2·159	·3712	2 $\frac{1}{2}$	7·854	4·908
$\frac{3}{4}$	2·356	·4417	3	9·424	7·068
$\frac{13}{16}$	2·552	·5814			

MEASUREMENT OF STONE-WORK.

1 Perch, Masons' or Quarrymen's Measure.

16½ feet long, 16 inches wide, 12 " high, } = 22 cubic feet. To be measured in wall.

16½ feet long, 18 inches wide, 12 " high, } = 24·75 cubic feet. To be measured in pile.

1 Cubic Yard = 3 feet × 3 feet × 3 feet = 27 cubic feet.

The cubic yard has become the standard for all contract work of late years.

Stone walls less than 16 inches thick count as if 16 inches thick to mason; over 16 inches thick, each inch additional is measured.

BRICKS REQUIRED FOR WALLS OF VARIOUS THICKNESS.

Number for Each Square Foot of Face of Wall.

Thickness of Wall.		Thickness of Wall.	
4 inch	7½	24 inch	45
8 "	15	28 "	52½
12 "	22½	32 "	60
16 "	30	36 "	67½
20 "	37½	42 "	75

Cubic Yard = 600 bricks in wall.
Perch (22 cubic feet) = 500 bricks in wall.
To pave 1 square yard on flat requires 41 bricks.
" " 1 " " edge " 68 "

GAS-HOLDERS.

A gas-holder 50 feet diameter, weighing 28,896 pounds, will rise with a pressure of 2·8 inches.

FORMULÆ FOR DETERMINING WEIGHT, PRESSURE, AND DIAMETER.

Let—

D = diameter;
P = pressure in inches of water;
W = weight of holder in pounds:

Then—

$$D^2 \times P \times 4{\cdot}128 = \text{weight in pounds.}$$

$$\frac{W}{D^2 \times 4{\cdot}128} = \text{pressure in inches of water.}$$

$$\sqrt{\frac{W}{P \times 4{\cdot}128}} = \text{diameter in feet}$$

Apply the above example:

$$W = 50^2 \times 2{\cdot}8 \times 4{\cdot}128 = 28{,}896 \text{ pounds.}$$

$$P = \frac{28{\cdot}896}{2500 \times 4.128} = 2{\cdot}8 \text{ inches.}$$

$$D = \sqrt{\frac{28{,}896}{2{\cdot}8 \times 4{\cdot}128}} = 50 \text{ feet.}$$

From diameter and pressure the first formula gives the weight.

From diameter and weight the second gives pressure.

From pressure and weight the third gives diameter.

The weight of a cubic foot of river or distilled water is 62·5 pounds = 1000 ounces, and gives this pressure on a square foot of surface.

A column of water one inch deep weighs 5·2 pounds, and gives this pressure on a square foot of surface.

RULE.—To ascertain the power necessary to overcome the friction of water through a horizontal pipe.

Cube the number of gallons to be discharged per second,

and multiply by the length of pipe in inches—divide the result by 140 times the 5th power of the diameter of the pipe in inches, the result is the number of H. Power.

Example.—What power is required to overcome the friction through a pipe 18 inches diameter, 3000 feet long, discharging 50 gallons per second. Ans. 17 H. P.

$$\frac{50^3 \times 3000 \times 12 = 4,500,000,000}{18^5 \times 140 = 264539520} = 17 \text{ H. P.}$$

Rule.—To find the Horse Power of a Steam Engine.

Multiply the area of piston in square inches by effective pressure in lbs., and multiply by the velocity of piston in feet per minute, and divide by 33000—$\frac{7}{10}$ of the result gives the Horse Power—the $\frac{3}{10}$ allowed for friction.

Rule.—To determine the *velocity of discharge*, of water through a pipe in feet per second.

Multiply 2500 times diameter in feet, by height in feet, divide by length in feet, added to 50 times the diameter—the square root of the quotient is the velocity in feet per second.

Rule.—To determine the quantity of discharge in cubic feet per second.

Multiply the area of the pipe in feet by the velocity in feet per second.

"Eytelwein's Formulas."

Rule.—To ascertain the height due to the friction of water through a pipe.

Square the velocity in feet per second, and multiply by the Co-eff. 13.88—and divide by diameter in inches, and multiply by length in feet, and divide by 2500.

Rule.—To determine the capacity of any size pipe.

Square the diameter in inches, and point off the right hand figure—$\frac{1}{3}$ of this sum is the content in ale gallons of each foot in length.

Hydraulic Mortar.

Lime, slaked	1 bushel.
Calcined Clay	1¼ "
Washed Sand	1¼ "

Concrete.

Unslaked Lime	3 bushels.
Sand	3 "
Gravel	2 "
Broken Stone	4 "

Cement.

Hydraulic Cement, 6 bushels, or 2 New York barrels.
Sand . . . 6 "

This will lay 1,000 bricks or two perches of stones.

Mortars.

Stone Lime (unslaked)	1 bushel.
Sand	3 "
Stone Lime (unslaked) . . .	1 "
Gravel	10 "

Cement for Leaks in Gas-holder.

Red Lead in oil	1 part.
White " "	1 "
Litharge "	1 "

Intimately mixed and pushed into crevices with wooden spatula.

For Leaks in Iron Retorts.

1. Fire-clay, 15 lbs.; Saleratus, 1 lb., made into paste with water. Applied inside to the broken part whilst the retort is at a good working heat; cover with fine coal dust and charge the retort for working.

For Outside.

2. Calcined fire-clay	$1\frac{1}{2}$ gallon.
Copper or brass filings	1 pint.
Gas tar (boiled)	3 gallons.

To be mixed, warmed, and worked into rolls about 1 inch thick.

Apply to the crack outside.

For Mending Iron Retorts.

3. 15 lbs. fire-clay, 1 lb. saleratus, with water to make a thick paste. This must be applied to the broken part of the retort, when at a good working heat; after this, cover it with a fine coal dust, and charge the retort for working.

To Stop Leaks in Clay Retorts when at working heat.

Five parts fire-clay.
Two " fine white sand.
One " (fused, and fine ground) borax.

Mix with water to the consistency of putty; take a lump of any convenient size, roll it in the hand to a proper length, and apply it over the crack, pressing upon it with a heavy force so as to drive it into the crack. Use a long spatula.

The borax fused as follows: Put an iron pot over the fire, into which drop pieces of borax, adding small portions at intervals, so as to prevent it flowing over the sides of the vessel; let it fuse until it becomes a glass in a liquid form, pour it off to cool, and then grind it fine.

This mixture should be kept on hand where clay retorts are used.

Cements for Cast-Iron.

1. One ounce of sal ammoniac to each hundred weight of borings, and use without heating.

2. In 100 lbs. of iron borings mix one ounce of flowers of sulphur, and add one pound of sal ammoniac, dissolved in hot water.

To preserve for use.—Put in an iron vessel, and cover with water.

To Remove the Deposit of Carbon from Clay Retorts.

Take out the stopper from stand pipe, so as to allow a current of air to pass through the retort and oxidize the carbon.

Use no bar. Put in a charge of coal after the retort has lain idle the number of hours required, and when it is withdrawn the carbon comes with it.

Iron Cement.

For securing mouth-pieces the iron cement is very valuable, or where heat is present.

1 ounce sal ammoniac.
1 " flowers of sulphur.
32 " clean cast-iron borings.

Mix well and keep dry.

When wanted for use mix into a paste with water, then apply to the joints and screw them together.

Or, as some prefer,

2 ounces sal ammoniac.
1 " flowers of sulphur.
16 " cast-iron borings.

Mix well and keep dry.

When wanted for use take of this 1 part, to which add 20 parts iron filings or borings; mix to a paste with water and apply to joints. These are improved by adding a little fine grindstone sand.

An Economic Cement for Mouth-Pieces.

1 part iron cement.
5 " fine clay.

Parts not exposed to heat may be joined with putty spread upon pasteboard, canvas, or woollen cloths.

Joints requiring to be separated often may be made of red and white lead mixed, and spread upon woollen cloths or canvas.

To Remove the Crust of Carbon from Clay Retorts.

Take out plug from stand-pipe, put a pipe through lid of mouth-piece, or leave a space open at bottom of lid to allow a current of cool air to pass, which oxidizes and crumbles the carbon, which is renewed with the following charge. A bar is injurious.

APPENDIX.

THROUGH the kindness of Henry P. M. Birkinbine, Chief Engineer of the Philadelphia Water Works, we are enabled to furnish the following :—

1 Rod of Brickwork = 272 sup. ft., 1½ bricks thick.
= 11⅓ cub. yds. = 306 cub. ft.
= 4000 bricks average work.
= 5300 " laid dry.

36 bricks flat, or 52 on edge = 1 yard paving.
No. of bricks in 1 cubic yd. = 384
1 load of Mortar = 1 cubic yard.
1 " " Sand = 1 " "
1 bag " Cement = 3 bushels.
1 sack " " = 5 "
1 cubic yard brickwork requires about { 6½ cubic foot sand. 2½ " " lime.

Mortar, Cement, &c.

Mortar.—1 of lime to 3 or 3½ of sharp river sand,
or 1 " 2 sand and 1 blacksmith's ashes, or coarsely ground coke.

Coarse Mortar.—1 of lime to 4 of coarse gravelly sand.

Concrete.—1 of lime to 4 of gravel and 2 of sand.

Hydraulic Mortar.—1 of blue lias lime to 2½ of burnt clay, ground together;
or 1 of blue lias lime to 6 of sharp sand, 1 of puzzolana, and 1 of calcined ironstone.

Beton.—1 of Hydraulic Mortar to 1½ of angular stones.

Cement.—1 of sand to 1 of cement. If great tenacity is required, the cement should be used without sand.

Waterproof Mastic Cement.—1 of red lead to 5 of ground lime and 5 of sharp sand, mixed with boiled oil;
or 1 of red lead to 5 of whiting and 10 of sharp sand, mixed with boiled oil.

Portland Cement is composed of clayey mud and chalk ground together, and afterwards calcined at a high temperature; after calcining it is ground to a fine powder.

Plastering.

	1 in. thick.	¾ in. thick.	½ in. thick.
1 bushel of cement, or 1.28 cub. feet will cover	1⅐ s. yd.	1½ s. yd.	2¼ s. yd.
1 cement and 1 sand	2¼ "	3 "	4½ "
1 cub. yd. of lime, 2 yds. of sand, and 3 bushels of hair will cover	75 yds. sup. render and set on brick, or 70 yds. on lath.		

Asphalte Flooring.

8 lbs. of asphalte composition will cover 1 sup. foot, ¾ inch thick.

Water Works.

1 gallon of water = 0.16 cub. ft. approximately.
1 cube ft. " = 6¼ gallons "

Consumption of water in towns:—

16 gallons per head per day in non-manufacturing towns.

20 " " " " manufacturing towns. The main should be large enough for double the average quantity.

Impounding reservoirs to contain above 120 days' supply in the rainy districts, and 200 days' supply in the less rainy districts of England.

Service reservoirs to contain 3 days' supply.

On the average about $\frac{6}{10}$ of the rain fall is available for storage.

Loss from overflow of storm-water about 10 per cent.

Evaporation 50 per cent. less on flat country than on an undulating rocky country.

Cast Iron Pipes—Pressure in.

Let H = Head of water, in feet.

P = Pressure of water, in lbs. per square inch.

D = Internal diameter of pipe, in inches.

t = Thickness of metal, in inches.

$P = 0.433\ H.$

$t = 0.000054\ H\ d + x.$

or $t = .000125\ P.d + x.$

$x = \frac{3}{8}$ inches for pipes, less than 12 inches diameter.

$= \frac{1}{2}$ " " " from 12 to 30 inches.

$= \frac{5}{8}$ " " " " 30 to 50 "

Rule for finding the Weight of Cast Iron Pipes :—

D = Diameter outside, in inches.

d = " inside, or bore, in inches.

W = Weight of 1 yard of pipe, in lbs.

$W = 7.35(D^2 - d^2).$

Let the diameter of a pipe = 3 inches bore, and let the metal be $\frac{1}{2}$ inch thick.

Then the outside diameter is 4 = D.

Then $W = 7.35(D^2 - d^2)$

$= 7.35(16 - 9)$

$= 7.35 \times 7$

$= 51.45$ = 1 yard of pipe in lbs.

$= 17.15$ = 1 foot " " " "

The weight of two flanges = about 1 foot of pipe.

Delivery of Water in Pipes.

D = Diameter of pipe, in inches.

H = Head of water, in feet.

L = Length of pipe, in feet.

W = Cubic feet of water discharged per minute.

$$W = 4.72 \frac{\sqrt{D^5}}{\sqrt{\frac{L}{H}}} \qquad D = .538 \sqrt[5]{\frac{L \times W^2}{H}}$$

Gas Works.

Each Lamp consumes 5 cubic feet per hour.

In winter each lamp consumes from 1800 to 2500 cubic feet per month.

In summer each lamp consumes from 1000 to 1800 cubic feet per month.

Average consumption for each lamp = 21000 cub. ft. per year.
Private burners about = 5000 " " " "

Motion of Gas in Pipes.

Q = Quantity of Gas, in cubic feet per hour.
L = Length of pipe, in yards.
D = Diameter of pipe, in inches.
H = Head of water pressure, in inches.
G = Specific gravity of Gas.

$$Q = 1350\, D^2 \sqrt{\frac{H\ D}{G\ L}}$$

$$D = .056 \sqrt[5]{\frac{Q^2\ G\ L}{H}}$$

G May be assumed = .42 for ordinary calculations.
H " " " = $\frac{1}{2}$ an inch to 1 inch.

Services for Lamps.

2	Lamps	40	feet	from main	require	$\frac{3}{8}$	bore	of	pipe.	Seldom use less than $\frac{3}{4}$ in.
4	"	40	"	"	"	$\frac{1}{2}$	"	"	"	
6	"	50	"	"	"	$\frac{5}{8}$	"	"	"	
10	"	100	"	"	"	$\frac{3}{4}$	"	"	"	
15	"	130	"	"	"	1	"	"	"	
20	"	150	"	"	"	$1\frac{1}{4}$	"	"	"	
25	"	180	"	"	"	$1\frac{3}{4}$	"	"	"	
30	"	200	"	"	"	$1\frac{1}{2}$	"	"	"	

Light.

Velocity of light 192,000 miles per second, nearly.

Decomposition of Light:—

Violet	=	maximum chemical ray.	
Indigo.			
Blue.			
Yellow	=	"	light ray.
Orange.			
Red	=	"	heat "

Sound.

Velocity of sound in air	=	1142 ft.	per	second.
" " " " water	=	4900 "	"	"
" " " " iron	=	17,500 "	"	"
" " " " copper	=	10,378 "	"	"
" " " " wood	=	12,000 "	"	"
to	=	16,000 "	"	"

Distance Sounds may be heard on a still day:—

Human voice	150	yards.
Rifle	5300	"
Military band	5200	"
Cannon	35,000	"

Specific Gravity.

To find the Specific Gravity of a Substance:—

W = weight of body in air.

X = " " " " water.

G = specific gravity.

$$G = \frac{W}{W - X}$$

If the substance be lighter than water, sink it by means of a heavier substance, and deduct weight of heavier substance.

To find the weight of a cubic foot of any substance when the specific gravity is given:—

Weight of a cubic foot, in lbs. $= 62.5 \times G$.

Rust Joint Cement (Quickly Setting).

1 Sal ammoniac in powder (by weight).
2 Flower of sulphur.
80 Iron borings made to a paste with water.

Rust Joint Cement (Slowly Setting).

2 Sal ammoniac.
1 Flower of sulphur.
200 Iron borings.

The latter cement is the best if the joint is not required for immediate use.

Red Lead Cement for Face Joints.

1 of white lead.
1 of red lead, mixed with linseed oil to the proper consistency.

Fuel.

Average Evaporative Power:—

1 lb. of	coke evaporates		9 lbs. of water.			
1 " "	coal	" (average)	9 "	"	"	13 have been in practice.
1 " "	slack	"	4 "	"	"	
1 " "	oak (dry)	"	$4\frac{1}{2}$ "	"	"	
1 " "	pine "	"	$2\frac{1}{2}$ "	"	"	

Coal loses about $\frac{1}{3}$ of its weight in coking, but increases in bulk $\frac{1}{10}$.

Stationary expansive condensing engines use from 5 to 7 lbs. of coal per horse power per hour.

Locomotive (passenger) from 25 to 30 lbs. per mile.
" (heavy goods) " 45 to 55 " " "

Wood-burning locomotives will run 24 miles with 1 cord of wood.

A cord of wood = 4 feet × 4 feet × 8 feet.

Navy allowance of stowage of coal = 2700 lbs.; 48 cubic feet per ton. The bulk of wood is about 6 times as much as an equivalent of coal.

Memorand. Connected with Water.

1 cubic foot of water = 62·4 lbs.
1 " inch = ·036 "
1 gallon = 10· " or 70,000 grs. imperial gal.
or = 0·16 cubic feet.
1 cubic foot of water = 6·2355 gallons.
or approximately = $6\frac{1}{4}$ "
1 cwt. of water = 1·8 cubic feet = 11·2 gals.
1 ton " = 35·9 " = 224· "

Pressure of Water per sq. inch at different heads:—
P = pressure in lbs. per square inch.
H = head of water in feet.
P = H × ·4333.
H = P × 2·31.
Pressure per square foot = H 62·4.
Cubic foot of water × 0·557 = cwt. approximately.
" " × 0·028 = tons "
1 cubic foot of sea water = 64.14 lbs.
Weight of sea water = weight of fresh water × 1·028.
1 gallon in wine, ale, or dry measure
= $277\frac{1}{4}$ cub. inches = ·16 cub. feet.
= 10 lbs. of distilled water.
Cubic feet × 6·232 = gallons English.
7.481 = " American.
" inches × ·003607 = "
1 bushel = 2218·19 cub. inches.
= 1·28 cub. feet.

Decimal Equivalents of Inches, Feet, and Yards.

Fractions of an inch.	Decms of an inch.	Decms of a Foot.
	.0625	= .00521
$\frac{1}{8}$	.125	= .01041
	.1875	= .01562
$\frac{1}{4}$	.25	= .02083
	.3125	= .02604
$\frac{3}{8}$	.375	= .03125
	.4375	= .03645
$\frac{1}{2}$	.5	= .04166
	.5625	= .04688
$\frac{5}{8}$	.625	= .05208
	.6875	= .05729
$\frac{3}{4}$	.75	= .06250
	.8125	= .06771
$\frac{7}{8}$	.375	= .07291
	.9375	= .07812
1 inch	1.00	= .08333

Inches.	Feet.	Yards.
1 =	.0833 =	.0277
2 =	.1666 =	.0555
3 =	.25 =	.0833
4 =	.3333 =	.1111
5 =	.4166 =	.1389
6 =	.5 =	.1666
7 =	.5833 =	.1944
8 =	.6666 =	.2222
9 =	.75 =	.25
10 =	.8333 =	.2778
11 =	.9166 =	.3055
12 =	1.000 =	.3333

Decimal Equivalents of lbs., qrs., and cwt.

qrs.	lbs.	cwt.	qrs.	lbs.	cwt.	qrs.	lbs.	cwt.	qrs.	lbs.	cwt.
0	0½	=.0044	1	0	=.25	2	0	=.5	3	0	=.75
0	1	.0089	1	1	.2589	2	1	.5089	3	1	.7589
0	2	.0178	1	2	.2678	2	2	.5178	3	2	.7678
0	3	.0268	1	3	.2768	2	3	.5268	3	3	.7768
0	4	.0357	1	4	.2857	2	4	.5357	3	4	.7857
0	5	.0446	1	5	.2946	2	5	.5446	3	5	.7946
0	6	.0535	1	6	.3035	2	6	.5535	3	6	.8035
0	7	.0625	1	7	.3125	2	7	.5625	3	7	.8125
0	8	.0714	1	8	.3214	2	8	.5714	3	8	.8214
0	9	.0803	1	9	.3303	2	9	.5803	3	9	.8303
0	10	.0892	1	10	.3392	2	10	.5892	3	10	.8392
0	11	.0982	1	11	.3482	2	11	.5982	3	11	.8482
0	12	.1071	1	12	.3571	2	12	.6077	3	12	.8571
0	13	.1160	1	13	.3660	2	13	.6160	3	13	.8660
0	14	.125	1	14	.375	2	14	.625	3	14	.875
0	15	.1339	1	15	.3839	2	15	.6339	3	15	.8839
0	16	.1429	1	16	.3929	2	16	.6429	3	16	.8929
0	17	.1518	1	17	.4018	2	17	.6518	3	17	.9018
0	18	.1607	1	18	.4107	2	18	.6607	3	18	.9107
0	19	.1696	1	19	.4196	2	19	.6696	3	19	.9196
0	20	.1786	1	20	.4286	2	20	.6786	3	20	.9286
0	21	.1875	1	21	.4375	2	21	.6875	3	21	.9375
0	22	.1964	1	22	.4464	2	22	.6964	3	22	.9464
0	23	.2054	1	23	.4554	2	23	.7054	3	23	.9554
0	24	.2143	1	24	.4643	2	24	.7143	3	24	.9643
0	25	.2232	1	25	.4732	2	25	.7232	3	25	.9732
0	26	.2321	1	26	.4821	2	26	.7321	3	26	.9821
0	27	.2411	1	27	.4911	2	27	.7411	3	27	.9911

Properties of the Circle.

Diameter	× 3·14159	=	circumference.
"	× ·8862	=	side of an equal square.
"	× ·7071	=	" inscribed "
Diameter²	× ·7854	=	area of circle.
Circumference	÷ 3·14159	=	diameter.
Radius	× 6·28318	=	circumference.
Circumference	= 3·54		$\sqrt{\text{area of circle}}$.
Diameter	= 1·128		$\sqrt{\text{area of circle}}$.

Mensuration of Surfaces.

Area of triangle = base × $\frac{1}{2}$ perpendicular.

Area of circle = diameter^2 × ·7854.

Area of sector of circle = arc × $\frac{1}{2}$ radius.

Area of sector of circle = $\frac{\text{No. of deg. in arc} \times \text{area of the circle.}}{360.}$

Area of parabola = base × $\frac{2}{3}$ height.

Area of ellipse = transverse axis × ·7854. Conjugate axis.

Area of cycloid = area of generating circle × 3.

Surface of cylinder = area of both ends + length × circumference.

Surface of cone = area of base + circumference of base × $\frac{1}{2}$ slant height.

Surface of sphere = diameter^2 × 3·1415 = diameter × circumference.

Surface of frustrum = sum of girt at both ends × $\frac{1}{2}$ slant height + area of both ends.

Mensuration of Solids.

Cylinder = area of one end × length.

Sphere = diameter^3 × 0·5236.

Segment of sphere = 0·5236 H ($H^2 + 3R^2$) where H = height of segment and R = radius of the base of segment.

Cone or pyramid = area of base × $\frac{1}{3}$ per perpendicular height.

Frustrum = $\frac{1}{3}$ H ($A + a + \sqrt{A \times a}$). When A and a = areas of the ends, H = perpendicular height.

Frustrum of Cone = 0·2618 H ($D^2 + d^2 + D . d$). When D and d = the diameters of each end and H = perpendicular height.

Wedge = area of base × $\frac{1}{2}$ perpendicular height.

Frustrum of wedge = $\frac{1}{2}$ H (A + a). When A and a = area at each end, H = perpendicular height.

Properties of Metals.

Metals.	W't of a cubic inch in lbs.	Specific Gravity.	Weight of a cubic foot in lbs.	Tenacity in lbs. per square inch.	Crushing force in lbs. per sq. inch.	Melting point. Fahr.	Expansion between 32° & 212°	Conducting power.
Aluminium . .	.092	2.56	160	...	...	[1]1800°	...	—
Antimony, cast .	.242	6.7	418	1066	...	810	.0011	—
Bismuth . . .	.35	9.82	615	3250	...	497	.0014	—
Brass, cast . .	.3	8.4	525	17,978	10,300	1800	.002	—
Brass, wire . .	...	8.5	531	49,000	...	...	...	—
Copper, cast . .	.32	8.89	555	19,072	11,700	1996	.0017	—
Copper, sheet .	...	8.95	559	33,000	...	...	...	898
Copper, wire .	...	9.	562	61,000	...	...	...	—
Gold	.7	19.25	1203	20,400	...	2016	.0016	1000
Gun metal . .	.3	8.4	525	36,000	...	...	...	—
Iron, wrought bar	.28	7.7	481	60,000	38,000	...	.0012	347
Iron, Swedish .	...	7.6	475	70,000	...	...	...	—
Iron, wire . .	...	...	...	85,000	...	...	...	—
Iron, cast . . .	.26	7.18	448	19,000	92,000	2786	·0011	—
Lead, cast . .	.41	11.35	709	1824	7,000	612	·0028	180
Lead, sheet . .	...	...	...	3328	...	...	...	—
Mercury . . .	.49	13.56	847	...	...	39	.016	—
Silver	.38	10.47	654	41,000	...	1873	.0019	973
Steel	.282	7.8	487	120,000	...	...	.0011	—
Steel, puddled .	...	7.78	485	80,000	...	...	...	—
Tin	.263	7.29	455	5000	15,000	442	.0021	304
Zinc	.253	7.	437	8000	...	773	.0029	363

[1] Approximate; no well authenticated experiments on aluminium.

Timber.	Specific gravity.	W't in lbs. per cub. ft.	Tenacity in lbs. per sq. inch.	Crushing force in lbs. per sq. in.
Ash	.8	50	17.200	9.000
Beech . . .	.69	43	11.000	9.000
Birch . . .	.71	44	15.000	5.500
Cedar . . .	.48	31	11.000	5.600
Deal, Christiana .	.7	44	12.000	6.000
Elm	.6	37	13.000	10.000
Hornbeam . .	.75	47	20.000	7.000
Larch . . .	.55	35	9.000	5.500
Memel . . .	.6	37		
Mahogany, Spanish .	.8	50	16.000	8.000
Oak, English . .	.93	58	17.000	10.000
Oak, Canadian . .	.87	54	10.000	6.000
Pine, red . . .	·65	41	12.000	5.800
Pine, yellow . .	·45	29	11.000	5.400
Teak, Moulmein .	·65	41	15.000	12.000
Yew	·8	50	8.000	
Miscellaneous.				
Asphaltum . .	.9	56		
Gutta percha . .	.98	61		
India rubber . .	.94	60		
Ivory	1.8	112		

Fluids.	Specific gravity.	Weight per cube foot.	Boiling point.	Expansion.[1]
Alcohol . . .	.8	50	173°	.11
Ether . . .	.74	46	100	.07
Oil	.90	56	...	.08
Water, fresh . .	1.000	62.4	212	.047
Water, sea . .	1.028	64.1	213	

Gases.	Water 1.	Weight (air being 1).	Weight of cube foot in grains.
Air	.0012	1.000	527
Carbonic acid . .	.0018	1.524	800
Carburetted hydrogen	.0005	.420	220
Hydrogen . . .	.00008	.069	43
Oxygen . . .	.00125	1.103	627

[1] Expansion of fluids is calculated between 32° and 212°.

Description.	Specific gravity.	Weight per cub. ft. in lbs.	Tenacity in lbs. per sq. inch.	Crushing force in lbs. per sq. inch.
Artificial Substances.				
Brick	2.0	124	290	1500
Brickwork, in mortar	1.6	100	50	
Brickwork, in cement	1.8	112 to 94	290	1000
Concrete, ordinary .	1.9	119		
Concrete, in cement .	2.2	133		
Cement, Portland .	1.3	81	290	1000
Cement, Roman .	1.	63		
Glass	2.5	156	9000	
Lime, quick . .	.8	50		
Mortar . . .	1.7	106	50	
Tile	1.8	112		
Stones, Earths, &c.				
Chalk	2.3	143	...	400
Clay	2.	125		
Coal	1.3	82		
Coke	.8	50		
Earth, rammed .	1.6	100		
Flint	2.6	163		
Gravel	1.9	120		
Granite	2.6	164	...	8,000
Grindstone . .	2.1	131		
Limestone . .	2.5	156	...	3000 to 8000
Marble	2.7	168	6000	6000
Sand	1.9	120		
Sandstone . .	2.5	156	...	5000
Stone, Bath . .	1.8	112		
Stone, Portland .	2.1	131		3700
York flag . . .	2.3	143	...	
Slate	2.8	175	9000	11,000
Shingle . . .	1.4	90		

Weight and Strength of Round Ropes of Hemp and Wire.

Hemp.		Iron Wire.		Steel Wire.		Equivalent Strength.	
Circumference.	lbs. weight per fathom.	Circumference.	lbs. weight per fathom.	Circumference.	lbs. weight per fathom.	Working load. per cwt.	Breaking strain. per ton.
2¾	2	1	1	...	...	6	2
—	—	1½	1½	1	1	9	3
3¼	4	1⅝	2	...	...	12	4
—	—	1¾	2½	1½	1½	15	5
4½	5	1⅞	3	...	...	18	6
—	—	2	3½	1⅝	2	21	7
5½	7	2⅛	4	1¾	2½	24	8
—	—	2¼	4½	...	...	27	9
6	9	2⅜	5	1⅞	3	30	10
—	—	2½	5½	...	...	33	11
6½	10	2⅝	6	2	3½	36	12
—	—	2¾	6½	2⅛	4	39	13
7	12	2⅞	7	2¼	4½	42	14
—	—	3	7½	...	...	45	15
7½	14	3⅛	8	2⅜	5	48	16
—	—	3¼	8½	...	...	51	17
8	16	3⅜	9	2½	5½	54	18
—	—	3½	10	2⅝	6	60	20
8½	18	3⅝	11	2¾	6½	66	22
—	—	3¾	12	...	...	72	24
9½	22	3⅞	13	3¼	8	78	26
10	26	4	14	...	...	84	28
—	—	4¼	15	3⅜	9	90	30
11	30	4⅜	16	...	...	96	32
—	—	4½	18	3½	10	108	36
12	34	4⅝	20	3¾	12	120	40

Weight and Strength of Flat Ropes of Hemp and Wire.

Hemp.		Iron Wire.		Steel Wire.		Equivalent strength.	
Size in inches.	lbs. weight per fathom.	Size in inches.	lbs. weight per fathom.	Size in inches.	lbs. weight per fathom.	Working load. per cwt.	Breaking strain. per ton.
$4 \times 1\frac{1}{8}$	20	$2\frac{1}{4} \times \frac{1}{2}$	11	...	...	44	20
$5 \times 1\frac{1}{4}$	24	$2\frac{1}{2} \times \frac{1}{2}$	13	...	...	52	23
$5\frac{1}{2} \times 1\frac{3}{8}$	26	$2\frac{3}{4} \times \frac{5}{8}$	15	...	...	60	27
$5\frac{3}{4} \times 1\frac{1}{3}$	28	$3 \times \frac{5}{8}$	16	$2 \times \frac{1}{2}$	10	64	28
$6 \times 1\frac{1}{2}$	30	$3\frac{1}{4} \times \frac{5}{8}$	18	$2\frac{1}{4} \times \frac{1}{2}$	11	72	32
$7 \times 1\frac{7}{8}$	36	$3\frac{1}{2} \times \frac{5}{8}$	20	...	...	80	36
$8\frac{1}{4} \times 2\frac{1}{8}$	40	$3\frac{3}{4} \times \frac{11}{16}$	22	$2\frac{1}{2} \times \frac{1}{2}$	13	88	40
$8\frac{1}{2} \times 2\frac{1}{4}$	45	$4 \times \frac{11}{16}$	25	$2\frac{3}{4} \times \frac{3}{8}$	15	100	45
$9 \times 2\frac{1}{2}$	50	$4\frac{1}{4} \times \frac{3}{4}$	28	$3 \times \frac{3}{8}$	16	112	50
$9\frac{1}{2} \times 2\frac{3}{8}$	55	$4\frac{1}{2} \times \frac{3}{4}$	32	$3\frac{1}{4} \times \frac{3}{8}$	18	128	56
$10 \times 2\frac{1}{2}$	60	$4\frac{5}{8} \times \frac{3}{4}$	34	$3\frac{1}{2} \times \frac{3}{8}$	20	136	60

Weight and Strength of Hemp Ropes.

W=Weight of rope, in lbs. per fathom.

C=Circumference, in inches.

B=Breaking weight, in tons.

$W = C^2 \times .26$.

$B = C^2 \times .28$ for hempen ropes.

$B = C^2 \times .2$ for common cables.

Rule for the Weight of Pipes.

D=Outside diameter of pipe in inches.

d=Inside diameter.

w=Weight of a lineal foot of pipe in lbs.

$w = k\ (D^2 - d^2)$.

k=2.45 for cast iron.

=2.64 for wrought iron.

=2.82 for brass.

=3.03 for copper.

=3.87 for lead.

Weight of a lineal foot of Cast-Iron Pipes in lbs.
2 flanges = 1 foot of length in weight.

Bore inch.	Thick-ness of metal.	Weight.	Bore inch.	Thick-ness of metal.	Weight.	Bore inch.	Thick-ness of metal.	Weight.
3	$\frac{3}{8}$	12.4	9	$\frac{3}{8}$	34.4	15	1	156.8
3	$\frac{1}{2}$	17.1	9	$\frac{1}{2}$	46.6	15	$1\frac{1}{8}$	177.7
3	$\frac{5}{8}$	22.2	9	$\frac{5}{8}$	58.9	16	$\frac{3}{4}$	123.1
4	$\frac{3}{8}$	16.1	9	$\frac{3}{4}$	71.7	16	$\frac{7}{8}$	144.7
4	$\frac{1}{2}$	22.1	10	$\frac{1}{2}$	51.4	16	1	166.6
4	$\frac{5}{8}$	28.3	10	$\frac{5}{8}$	65.1	16	$1\frac{1}{8}$	188.7
5	$\frac{3}{8}$	19.8	10	$\frac{3}{4}$	79.0	18	$\frac{3}{4}$	137.9
5	$\frac{1}{2}$	26.9	10	$\frac{7}{8}$	93.3	18	$\frac{7}{8}$	161.8
5	$\frac{5}{8}$	34.4	11	$\frac{1}{2}$	56.4	18	1	186.2
5	$\frac{3}{4}$	42.3	11	$\frac{5}{8}$	71.0	18	$1\frac{1}{8}$	210.8
6	$\frac{3}{8}$	23.4	11	$\frac{3}{4}$	86.4	20	$\frac{7}{8}$	178.9
6	$\frac{1}{2}$	31.9	11	$\frac{7}{8}$	101.8	20	1	205.8
6	$\frac{5}{8}$	40.6	12	$\frac{5}{8}$	77.3	20	$1\frac{1}{8}$	232.9
6	$\frac{3}{4}$	49.7	12	$\frac{3}{4}$	93.7	20	$1\frac{1}{4}$	260.3
7	$\frac{3}{8}$	27.1	12	$\frac{7}{8}$	110.4	22	1	225.4
7	$\frac{1}{2}$	36.8	12	1	127.4	22	$1\frac{1}{8}$	254.9
7	$\frac{5}{8}$	46.7	14	$\frac{5}{8}$	89.6	22	$1\frac{1}{4}$	284.8
7	$\frac{3}{4}$	56.8	14	$\frac{3}{4}$	108.4	24	1	245.0
8	$\frac{3}{8}$	30.8	14	$\frac{7}{8}$	127.5	24	$1\frac{1}{8}$	276.9
8	$\frac{1}{2}$	41.6	14	1	147.0	24	$1\frac{1}{4}$	319.3
8	$\frac{5}{8}$	52.8	15	$\frac{3}{4}$	115.7			
8	$\frac{3}{4}$	64.3	15	$\frac{7}{8}$	136.1			

Weight of Lead Pipes in lbs. per foot lineal.

Bore inch.	Common. lbs.	Middling. lbs.	Strong. lbs.
½	1.07	—	—
¾	1.6	1.8	2.
1	2.0	2.6	2.8
1¼	3.0	3.7	4.4
1½	4.0	4.7	5.6
2	5.0	6.0	7.0
2½	7.0	8.6	10.0

Weight of Copper Pipes in lbs. per foot lineal.

Bore inch.	Thickness in parts of inch.			
	1/16	⅛	3/16	¼
½	.42	.94	1.60	2.27
¾	.62	1.33	2.17	3.02
1	.79	1.69	2.66	3.77
1½	1.15	2.44	3.85	5.30
2	1.55	3.21	5.00	6.80
2½	1.94	3.97	6.13	8.31
	2.3	4.73	7.24	9.84

Birmingham Wire Gauge, compared with inches.

B.W.G.	= ins.	B.W.G.	= ins.	B.W.G.	= ins.	B.W.G.	= ins.
No. 1	= .31	No. 10	= .137	No. 19	= .042	No. 28	= .014
2	.28	11	.125	20	.035	29	.013
3	.26	12	.109	21	.032	30	.012
4	.24	13	.095	22	.028	31	.01
5	.22	14	.083	23	.025	32	.009
6	.2	15	.072	24	.022	33	.008
7	.187	16	.065	25	.02	34	.007
8	.166	17	.056	26	.018	35	.005
9	.158	18	.049	27	.016	36	.004

Weight of a Lineal Foot of Flat Bar Iron in Pounds.

Breadth in inches.	Thickness in fractions of inches.								
	1/4	5/16	3/8	7/16	1/2	5/8	3/4	7/8	1
1	.83	1.04	1.25	1.46	1.67	2.08	2.50	2.92	3.34
1 1/8	.93	1.17	1.40	1.64	1.87	2.34	2.81	3.28	3.75
1 1/4	1.04	1.30	1.56	1.82	2.08	2.60	3.13	3.65	4.17
1 3/8	1.14	1.43	1.72	2.00	2.29	2.87	3.44	4.01	4.59
1 1/2	1.25	1.56	1.87	2.19	2.50	3.13	3.75	4.38	5.00
1 5/8	1.35	1.69	2.03	2.37	2.71	3.39	4.07	4.70	5.43
1 3/4	1.46	1.82	2.19	2.55	2.92	3.65	4.38	5.11	5.84
1 7/8	1.56	1.95	2.34	2.74	3.13	3.91	4.69	5.47	6.26
2	1.67	2.08	2.50	2.92	3.34	4.17	5.01	5.86	6.68
2 1/8	1.77	2.21	2.66	3.10	3.55	4.43	5.32	6.21	7.10
2 1/4	1.87	2.34	2.81	3.28	3.76	4.69	5.63	6.57	7.52
2 3/8	1.98	2.47	2.97	3.47	3.96	4.95	5.95	6.94	7.93
2 1/2	2.08	2.60	3.13	3.65	4.17	5.21	6.26	7.30	8.35
2 5/8	2.19	2.74	3.28	3.83	4.38	5.47	6.57	7.67	8.77
2 3/4	2.29	2.87	3.44	4.01	4.59	5.74	6.88	8.03	9.18
2 7/8	2.40	3.00	3.60	4.20	4.80	6.00	7.20	8.40	9.60
3	2.50	3.13	3.75	4.38	5.01	6.26	7.51	8.76	10.02
3 1/4	2.71	3.39	4.07	4.74	5.43	6.78	8.14	9.49	10.86
3 1/2	2.92	3.65	4.38	5.11	5.84	7.30	8.76	10.23	11.69
3 3/4	3.13	3.91	4.68	5.47	6.26	7.82	9.39	10.95	12.52
4	3.34	4.17	5.00	5.84	6.68	8.35	10.02	11.69	13.36
4 1/4	3.54	4.43	5.32	6.21	7.09	8.87	10.64	12.42	14.19
4 1/2	3.75	4.69	5.63	6.57	7.51	9.39	11.27	13.15	15.03
4 3/4	3.06	4.95	5.94	6.94	7.93	9.91	11.89	13.88	15.86
5	4.17	5.21	6.26	7.30	8.35	10.44	12.52	14.61	16.70
5 1/4	4.38	5.47	6.57	7.67	8.76	10.96	13.14	15.34	17.53
5 1/2	4.59	5.73	6.88	8.03	9.18	11.48	13.77	16.07	18.37
5 3/4	4.80	6.00	7.20	8.40	9.60	12.00	14.40	16.80	19.20
6	5.01	6.25	7.51	8.76	10.02	12.53	15.03	17.53	20.05

Weight of a Lineal Foot of Round and Square Bar Iron, in lbs.

Diam. or side.	Square bars.	Round bars.	Breadth or diam. in ins.	Square bars.	Round bars.	Breadth or diam. in ins.	Square bars.	Round bars.
1/4	.209	.164	1	3.34	2.62	2 7/8	27.61	21.68
5/16	.326	.256	1 1/8	4.22	3.32	3	30.07	23.60
3/8	.470	.369	1 1/4	5.25	4.09	3 1/4	35.28	27.70
7/16	.640	.502	1 3/8	6.35	4.96	3 1/2	40.91	32.13
1/2	.835	.656	1 1/2	7.51	5.90	3 3/4	46.97	36.89
9/16	1.057	.831	1 5/8	8.82	6.92	4	53.44	41.97
5/8	1.305	1.025	1 3/4	10.29	8.03	4 1/4	60.32	47.38
11/16	1.579	1.241	1 7/8	11.74	9.22	4 1/2	67.63	53.12
3/4	1.879	1.476	2	13.36	10.49	4 3/4	75.35	59.18
13/16	2.205	1.732	2 1/8	15.08	11.84	5	83.51	65.58
7/8	2.556	2.011	2 1/4	16.91	13.27	5 1/4	92.46	72.30
15/16	2.936	2.306	2 3/8	18.84	14.79	5 1/2	101.03	79.35
			2 1/2	20.87	16.39	5 3/4	114.43	86.73
			2 5/8	23.11	18.07	6	120.24	94.43
			2 3/4	25.26	19.84			

To convert into weight of other metals, multiply tabular No. for cast iron by .93, for steel × 1.01, for copper, × 1.15, for brass × 1.09, for lead × 1.48, for zinc × .92.

Weight of Nuts and Bolt Heads, in lbs.

Diameter of bolt in inches	1/4	3/8	1/2	5/8	3/4	
Weight of hexagon nut and head	.017	.057	.128	.267	.43	.73
Weight of square nut and head	.021	.069	.164	.320	.55	.88

Diameter of bolt in inches	1	1 1/4	1 1/2	1 3/4	2	2 1/2	3
Weight of hexagon nut and head .	1.10	2.14	3.78	5.6	8.75	17	28.8
Weight of square nut and head .	1.31	2.56	4.42	7.0	10.5	21	36.4

Weight of a Square Foot of Sheet Metals in lbs.—Thickness Birmingham Wire Gauge.

Thickness B.W.G.	Iron.	Copper.	Brass.	Thickness B.W.G.	Iron.	Copper.	Brass.
30	.5	.58	.55	15	2.82	3.27	3.10
29	.56	.64	.61	14	3.12	3.60	3.43
28	.64	.74	.70	13	3.75	4.34	4.12
27	.72	.83	.79	12	4.38	5.08	4.81
26	.80	.92	.88	11	5.00	5.80	5.50
25	.90	1.04	.99	10	5.62	6.50	6.18
24	1.00	1.16	1.10	9	6.24	7.20	6.86
23	1.12	1.30	1.23	8	6.86	7.90	7.54
22	1.25	1.45	1.37	7	7.50	8.70	8.25
21	1.40	1.62	1.54	6	8.12	9.40	8.93
20	1.54	1.78	1.69	5	8.74	10.10	9.61
19	1.70	1.97	1.87	4	10.00	11.60	11.00
18	1.86	2.15	2.04	3	11.00	12.75	12.10
17	2.18	2.52	2.40	2	12.00	13.90	13.10
16	2.50	2.90	2.75	1	12.50	14.50	13.75

Weight of a Superficial Foot of Plates, Different Metals, in lbs.

Thick. inches.	Iron.	Brass.	Copper.	Lead.	Zinc.	Thickness.
$\frac{1}{16}$	2.5	2.7	2.9	3.7	2.3	.0625 in.=16 B.W.G
$\frac{1}{8}$	5.	5.5	5.8	7.4	4.7	.125 " =11 "
$\frac{3}{16}$	7.5	8.2	8.7	11.1	7.0	.1875 " = 7 "
$\frac{1}{4}$	10.	11.0	11.6	14.8	9.4	.25 " = 4 "
$\frac{5}{16}$	12.5	13.7	14.5	18.5	11.7	.3125 " = 1 "
$\frac{3}{8}$	15.	16.4	17.2	22.2	14.0	.375
$\frac{7}{16}$	17.5	19.2	20.0	25.9	16.4	.4375
$\frac{1}{2}$	20.0	21.9	22.9	29.5	18.7	.5
$\frac{9}{16}$	22.5	24.6	25.7	33.2	21.1	.5625
$\frac{5}{8}$	25.	27.4	28.6	36.9	23.4	.625
$\frac{11}{16}$	27.5	30.1	31.4	40.6	25.7	.6875
$\frac{3}{4}$	30.	32.9	34.3	44.3	28.1	.75
$\frac{13}{16}$	32.5	35.6	37.2	48.0	30.4	.8125
$\frac{7}{8}$	35.	38.3	40.0	51.7	32.8	.875
$\frac{15}{16}$	37.5	41.2	42.9	55.4	35.1	.9375
1	40.	43.9	45.8	59.1	37.5	1.000

Weight of Cast-Iron Balls and Solid Cylinders, in Pounds.

Diameter in inches,	1	2	3	4	5	6	7	8	9
Cast-iron balls,	.136	1.10	3.70	8.7	17.1	29.5	47.	70.	100.
Cast-iron cylinder 1 foot long	2.4	9.9	21.9	39.0	61.0	89.0	120	156	198

Weight of Ordinary Angle Iron, in lbs. per lineal foot.

Breadth, in inches,	1¼	1½	1¾	2	2¼	2½	2¾	3	3¼	3½
Weight per foot, in lbs.	1.8	2.7	3.3	3.9	5	6.5	8.3	10.4	11.7	14.

Strength and Weight of Chains.

Chains.			Chain cables.		
Diameter in inches.	Weight per fathom in lbs.	Proof strain in cwt.	Diameter in inches.	Weight per fathom in lbs.	Proof strain in cwt.
$\frac{5}{16}$	5½	25½	½	13¾	80
⅜	8	36¾	⅝	22	120
½	14	65½	¾	30	200
⅝	22	102	⅞	42	260
¾	32	147	1	55	360
⅞	43	200	1⅛	68	520
1	56	268	1¼	84	640
1⅛	71	334	1⅜	102	760
1¼	87	408	1½	120	880
1⅜	106	498	1¾	148	1200
			2	180	1600

CIRCULAR

TO GAS LIGHT COMPANIES AND GAS ENGINEERS.

HARRIS & BROTHER,

GAS METER MANUFACTURERS,

No. 1117 Cherry Street,

PHILADELPHIA.

Among the many, we subjoin a few of our

CERTIFICATES OF RECOMMENDATION.

PHILADELPHIA GAS WORKS, *February* 14, 1857.

MESSRS. HARRIS & BROS.

DEAR SIRS: I take great pleasure in stating that we have found your meters to register with entire correctness, and to be generally of very good quality. The number furnished by you to these works is over 5,000.

Very truly,

JOHN C. CRESSON,

Engineer.

OFFICE CAMDEN GAS COMPANY,
CAMDEN, N. J., *March* 4, 1858.

MESSRS. HARRIS & BROS.

DEAR SIRS: In answering your inquiry, I will say that your meters first came under my notice in 1850. At that time I was Superintendent of the Northern Liberties Gas

Company; and as there were a large number of manufacturers urging their meters for our patronage, I instituted a thorough investigation of their various claims to our confidence for correct registration and mechanical construction. The investigation satisfied me of the good quality of your meters, so much so that a large portion of those purchased for the Company while I was with them, were from your manufactory.

We have your meters in use here in Camden now that have been in use about five years, and give entire satisfaction. In fact, I can say with pleasure that I consider your meters in all respects as good as any that have come under my notice.

Yours, truly,

O. W. GOODWIN,

Superintendent Camden Gas Company.

PHILADELPHIA, *March* 11, 1858.

THIS IS TO CERTIFY, that we have, since they commenced business, mainly furnished Harris & Brothers with the materials which they consume in the manufacture of meters, and it affords us pleasure to state that in every sale made them, quality was the first consideration, they invariably insisting upon having none other than the choicest brand of tin plate and the primest quality of block tin; and for the first named articles we have imported for them a very heavy and exceedingly high cost article where strength of material is required in the manufacture of their meters.

NATHAN TROTTER & CO.,

No. 36 *North Front Street.*

OFFICE OF THE GAS LIGHT COMPANY OF AUGUSTA, GA.,
March 6, 1858.

MESSRS. HARRIS & BROS.

GENTS: Yours of the 1st inst. has just reached me. I cheerfully comply with your request.

This is to certify that having used meters manufactured by Messrs. Harris & Bros. for the last nine years, we do not hesitate to state they have given entire satisfaction in point of workmanship and accuracy of registration; they are surpassed by none, and equalled by few manufactories in the Union. We have used meters made by other establishments, but our preference is in favor of those made by Harris & Bros. of Philadelphia, Pennsylvania.

G. S. HOOKEY,
Sup't of the Gas Light Company of Augusta, Ga.

PHILADELPHIA, *March* 4, 1858.

To HARRIS & BROS.

GENTLEMEN: In regard to the questions you ask me of my opinion respecting your meters, I can only answer as far as my practical experience in the business has taught me, having worked for Code, Hopper & Co. some six years, and yourselves about two years; my position has been such as to acquire the knowledge of the different mixtures of the metals used in the manufacture of meters and the quality of material. I most honestly believe your meter to be equal, if not better than any manufactured in this city. I have also visited several Gas Works where your meters were in use, and upon inquiry, found that they give entire satisfaction.

Yours, respectfully,
DANIEL P. VANDEGRIFT,
1508 *Filbert Street.*

March 14, 1857.

MESSRS. HARRIS & BROS.

GENTLEMEN: For the last ten years we have used in the Philadelphia Gas Works meters made in your factory, and take pleasure in expressing my opinion as to their correctness. On several occasions I have taken your meters apart, and found the materials used in the manufacture of the same to be of as fine a quality as is used in those imported from Europe, or made by any of the manufacturers in the United States.

Respectfully yours,

GEORGE WIEGAND,

Inspector of Gas Fittings, Philadelphia Gas Works.

October, 1864.

MESSRS. HARRIS & BROS.

Your Gas Meters were first used by me in 1852, when building the Gas Works in Germantown, now 22d Ward of this city. Many of these meters are still in use and in good working order. This is the best evidence and the highest testimony of their durability. In 1855 I ordered 250 for the Gas Company in West Philadelphia, now the 24th Ward. The Gas Company demanded their inspection by, and the approval of, the Chief Engineer of the Philadelphia Gas Works, Jno. C. Cresson, Esq.

That none of the 250 should be rejected, neither running too fast nor too slow, became a subject of remark among the Provers—so much for their accuracy. To this nothing need be added. Your late invitation to examine the materials used in your establishment satisfies me that you still maintain the same high character of your meters.

From a personal acquaintance with the members of your firm, I can commend them to the confidence and patronage of Gas Companies.

Yours, respectfully,

H. P. M. BIRKINBINE,

Chief Engineer Philadelphia Water Works.

We manufacture and keep constantly on hand the following articles, viz :—

Wet and Dry Consumer's Meters; Station, Experimental, and Glazed Meters; Pressure Registers, Indicators, and Gauges; Meter Provers and Photometers; Governor Drums and Centre Seal Drums.

We furnish all articles appertaining to Gas Works.

Our long experience and practical knowledge, and the constant supervision of our factory guarantee durability, accuracy, and excellence of workmanship.

Orders for any articles not in our line, which Gas Engineers or Superintendents of Gas Works may want, will be furnished with pleasure without any additional cost or expense.

We respectfully solicit your patronage, and think you will agree with us that the interests of Gas Companies will be better served by keeping up a competition in the trade, than by allowing its monopoly.

Prompt attention given to all orders addressed to

HARRIS & BROTHER,
Gas Meter Manuf'rs, No. 1117 *Cherry St., Phila.*

www.ingramcontent.com/pod-product-compliance
Lightning Source LLC
LaVergne TN
LVHW021430110826
845150LV00007B/2171

9781425507046